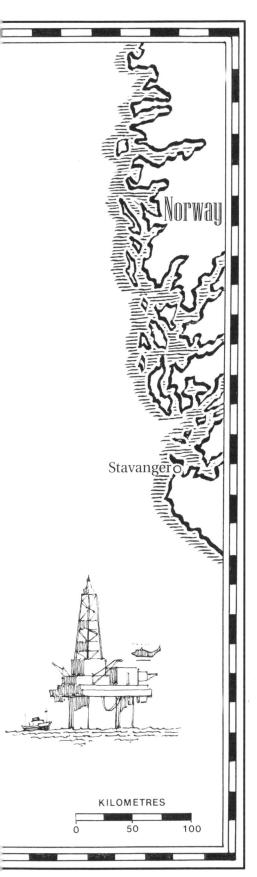

Key to the Oil Fields

D1186588

3. Eider

4. Tern

5. Murchison

6. Statfjord

7. Dunlin

8. North Cormorant

9. South Cormorant

10. Hutton

11. Brent

12. North West Hutton

13. Heather

14. Ninian

15. North Alwyn

16. Alwyn

17. Beryl

18. Piper

19. Claymore

20. Tartan

21. Beatrice

22. Maureen

23. Buchan

24. Forties

25. Montrose

26. Fulmar

27. Auk

JUST BEING THERE

With Bears and Tigers in the North Sea

JUST BEING THERE

With Bears and Tigers in the North Sea

Andrew Wylie

Richard:

The master of syntax ... and
friend and colleague.

Affectionately

Andrew.

Andrew.

16.02.06.

DUNEDIN ACADEMIC PRESS

EDINBURGH

Published by
Dunedin Academic Press Ltd
Hudson House
8 Albany Street
Edinburgh EH1 3QB
Scotland

ISBN 1 903765 41 2

BRITISH LIBRARY CATALOGUING IN PUBLICATION DATA
A catalogue record for this book is available from the British Library

Composed in 10¼/14¼pt Utopia with Akzidenz Grotesque display
by Makar Publishing Production, Edinburgh

Printed and bound in Great Britain by Cromwell Press

Contents

Acknowledgements

This sort of book is the result of a lot of hard work by many people from the very outset of the Oil Chaplaincy.

Standing alone as the mistress of networking and quiet guidance, was my former secretary Anne Fullerton. Then there were the 'door openers' and encouragers, and behind the scenes workers at every level of responsibility and from many companies: Douglas Allsop, Andrew Armstrong, Hayden Barrett, Ronnie Compton, Jim Farquherson, Ian Henderson, David Jamieson, Sam Laidlaw, John Moorhouse, John Robertson, Peter Ryalls, Joyce Simpson and many, many more. In the area of government the concerned interest of Lord Moynihan and the late Sir Peter Morrison meant a great deal.

My gratitude for support in what could be a lonely job starts with the Norwegian offshore team: Terje Bjerkholt, Kare Mjolhus, Knot Milbach, Petter Skants and Arvid Tweit. We shared much together. Closer to home, the late Sister Mary Macdonald, and my professional colleagues Donald Rennie, Donald Ross and James Stewart were staunch and determined supporters. Douglas Anderson, Linda Cameron, Professor Ray Furness, Mike Marshall, Geoffrey Ruddock and Dick Wrathall prodded, urged, criticised and helped in many practical ways. Pamela Ritchie then translated everything into order and put it into manuscript form. Dr Douglas Grant and Anthony Kinahan at Dunedin Academic Press, with much patience and kindness, brought everything to pass. My gratitude to Sir Bob Reid is great, for his preparedness to trust me.

I am blessed with a soulmate; Jennifer ('wee blondie') is the best sort of wife, providing trenchant criticism and hours of practical help.

Finally, my thanks to Guinness and Mungo for their loyal companionship on my many 'thoughtful' walks.

Foreword

The North Sea has carried ships large and small in peace and in war, sometimes above the stormy waves and sometimes below, but all going somewhere. Now, in the middle of that sea, braving its Arctic winds, stand huge structures going nowhere, pitting the strengths of their steel against the unrelenting power of the ocean.

The story of these islands of technology is a story about human endeavour, about leadership, about discipline and about brains and expertise. It is simply about people contributing their lives and their loyalty, knowing as they stand on these islands that it is not just about money. It is something more and the sea in its moods makes it something much deeper, something more fundamental.

For an engineer to write the story would be too mundane, too defined to catch the ephemeral. For a clergyman to write – and one who had lived and listened to it – is a unique opportunity to feel the spirit and to access the thoughts of the individual in the face of overwhelming elements.

This is not a religious book. It is a book about people rich in their physical achievement but even wealthier in their thoughts and their contemplations. Andrew Wylie has the humanity and the empathy to write a book that needed to be written. I sincerely thank him.

Rob Reid

Introduction

My admiration and respect for those who work offshore is immense and I wanted to write about these men *and* women for two distinct reasons. Firstly, to pay tribute to what must be the most unacknowledged workforce in British industrial history. I think this lack of recognition stems from the wide geographical area from which it is drawn: on the mainland, what oilworkers call the beach, there are no identifiable communities of offshore workers and because their workplace is over the horizon, the move from being out of sight to out of mind is inevitable. The second reason for the book is to attempt to portray a significant chapter in Britain's social and industrial history: one which reveals the adaptability of men (and it was mostly men), when they forsook landlocked jobs in industries that all too often, sadly, were in terminal decline, and relocated in a maritime environment.

It will be all too evident that my engineering background is zero and the disciplines of that specialist activity remain a closed book to me. Perhaps, surprisingly, I believe my ignorance to have been an advantage because I surveyed the whole offshore scene through unconditioned eyes. My sole concern was for the men and women who went about their business in an unusual location; coping with a demanding job, a relentless and predictable routine, and endured a grindingly regular separation from their loved ones. Setting a clergyman down in the unique offshore location can seem both odd and rather pointless. I have tried to show that it was neither. It has to be said that this conviction became stronger as the book was being put together. The incidental development, and clarification, of my own thinking in those areas with which the Church wrestles at this time, become evident as the pages unfold.

My own, somewhat idiosyncratic, path as a cleric has been dictated by a strong personal vision. There have been times when I have been led up disconcerting cul de sacs. To my surprise, the North Sea was the location where everything, for once, came together.

There will be those who can happily give chapters 2 and 3 a miss. They are not about the offshore worker – the reason for these pages – but about myself! The sole justification for this apparent ego trip is that I felt it necessary to detail those experiences – both good and bad – in my own life that make me convinced that all that went before was a preparation the better to fit me for the offshore venture. Similarly, the personal musing in chapter 14 is included as my considered apologia for the validity of the cleric in the midst.

I kept a detailed log of my offshore trips, which was written up immediately on my return to the beach. The brief excerpts at the start of each chapter, which are not in chronological order, should help to prepare the way for what follows. Like every other workforce, those offshore have their own (repeatable!) vocabulary – the Glossary at the end of the book could prove useful.

As the first chaplain to the offshore oil industry, I was afforded a unique, and privileged, overview of the whole of North Sea activity – one that has been shared only by my long-serving successor, Angus Smith. Of course there are changes in offshore operation but the basic purpose and methodology remain the same.

The book is not a success story of a questing individual who sought to express the love of God. Rather it charts an erratic course with its disappointments and failure, as well as the points of purpose and direction. For all that, all the time – shining through everything – was the resilience and generosity of the human spirit. Thank you, Bears and Tigers!

1

The New Boy

'No-one like you has
ever been here before.'
Oil office receptionist

'Sit down and tell me in
ten minutes why the
Church should believe
it has something to
offer the oil industry.'
CEO of a large oil company

The winter of 1985 was a cold one, even by Aberdeen's standards. The snow and slush leaked through my shoes as I plodded from one household name in the petroleum world to the next. It dawned on me that it was going to be difficult enough to access the decision-makers, let alone talk to them. Office security can be a bit of a double whammy. Not only does it stop the entry of undesirables but it keeps out those who are uncertain as to whom they want to meet. Breezing into a building, knocking on a door, and securing an interview is a fantasy that belongs to the 1950s. Doing your homework and getting hold of the right names (not always easy) is but the first step and there is no room for mistakes. You do not get a second chance.

Early on, I tripped up with one oil company, when going through my increasingly well-thumbed oil industry 'Who's Who' and falling for a very grand title which suggested its holder would be able to open all doors. I could not have been more wrong. The title was unduly aspiring and the individual in question had no real decision-making powers. What he was able to do, as I was to discover, was to stymie all future approaches. All the time he made out

1

that he was dealing with my expressed wish to visit the offshore workforce he was creating irritating obstacles.

As in all walks of life, I learned the hard way and came to understand that, no matter how strong your advocacy, if you did not make your move at the right level then the whole exercise becomes a waste of time. Very quickly I realised that I had to work from the top, where the carpets were thickest and the crucial decisions were made.

To arrange a meeting with the Chief Executive Officer of a major international oil company was not easy. I developed a reluctant admiration for their personal assistants who, without exception, appeared to have a total commitment to the protection of their bosses from anything that might be deemed an irrelevant intrusion. Quite by accident I discovered a mild stratagem that worked. One day, when seeking an audience, I described myself to the PA as 'a minister'. I did not think of adding 'of religion', and that was the last thing that occurred to the PA. Ministers of State and of foreign governments were all keeping a close interest in the UK's oil strategies, and sudden visits were not unknown. It was all a matter of interpretation and I found myself the beneficiary.

Only one company made it very plain, from the outset, that their policy forbad involvement with religious institutions of any sort. Even then I always had the impression that the Chaplaincy was put into convenient cold storage where it could be pulled out in the event of untoward emergency.

When, at last, I reached the inner sanctum, it was all rather like Beijing's 'Forbidden City': as you passed through one reception area after another until at last you stood before Himself. Without exception, I was received with immaculate courtesy, albeit laced with a scarcely concealed curiosity. You could sense a mind ticking over and asking just what on earth does this Reverend have to say to me?

The setting was invariably the same. I would find myself facing a large man (why are so many men in the oil industry size XXL?). He would be seated in a large chair behind a large desk. Preliminary courtesies were dispensed with pretty quickly, and then would come the crucial question: 'Reverend Wylie, just what is the contribution the Church can make to the operation of the offshore oil industry?' At least it was the reverse of the usual question, and it was perfectly right and proper in an industry, which, like others, had to concentrate on the technical and financial.

It was not an easy question to answer simply. To begin with I made a mess of my response, pleading for a place for the things of the Spirit in the world of work. I would get increasingly convoluted and would go on for too long.

Poor CEO, it must have been a nightmare for him, which he longed to come to an end. But, over the weeks my presentation improved – it had to. On one occasion I rambled away and then said, 'I am not making a very good job of this,' to which came the encouraging response, 'Nope!'

Eventually I grasped that the key question, as yet unasked, was not to do with spirituality but overt religiosity. I realised that managers could live with the former – even before coffee on a Monday morning – but there was a real concern that the staff might be 'got at'. I set out to reassure these busy men, who were generous to me with their valuable time, that I was not in the business of proselytising. In every way possible, I sought to emphasise that the Church wanted to offer support and show its concern, and God's love, for those who worked both on the beach and offshore.

I think the reason for my inadequate advocacy for a chaplain was explained by my professional life, most of which had been spent in situations where either the clergyman was in clear control, or where the presence of the clergy was accepted without question. If there was to be the remotest chance of a credible North Sea chaplaincy, I had to start by getting my act together and securing the trust and confidence of those at the very top.

It was all too easy for the Church to say it had a spiritual responsibility for those offshore, but converting a bland statement into a reality was going to take time. I knew I was asking a great deal of the operators. At the most practical level a flight in a helicopter was not cheap and bed and board on a platform came at a price. It was not difficult to understand why companies were cautious about the whole project. Common sense suggested that only by hastening slowly would a favourable decision be reached. I felt like a batsman settling down to play a long innings.

The oil industry can be identified by its pragmatism, and I was seeking an answer to a request for which there was no precedent, which provided no bottom lines, and where positive benefit would be hard to assess. And whilst companies were coming to their conclusions, both individually and corporately, I found it fascinating to glimpse the operation of very different management styles.

As the outsider it seemed to me they were reflected in the office furnishings. There was the chrome and leather brigade who positively breathed an antiseptic efficiency. In vivid contrast were those who favoured mahogany desks and wood panelling – evocative of long established, sound and solid performance. Some offices were particularly lived-in and there was one where I would invariably be greeted by the strains of Vivaldi or Delius as I entered the room.

There were office blocks that brought back memories of Milan railway station after Mussolini had rebuilt it to reflect the glories of the Roman Empire. This style was in almost absurd contrast to the companies that sought to insert their employees into small single offices in long corridors with identical doors – rather like the housing of battery hens.

I became a connoisseur of canteen food. Provision for the inner man was surprisingly variable. In the best Michelin tradition, I would have awarded a star to Total. The French gastronomic tradition was upheld and could do much to lighten a grey January day in Aberdeen. In the main, the canteens deserved to be called restaurants, with food that was either free or generously subsidised. The reaction of staff to this perk intrigued me. Maybe we all take so much for granted, but why was it that the higher the standard of food the greater the grumbles?

In, for me, this strange new world I found myself making the most unlikely comparisons. It was startling to note that the proceedings of the World Council of Churches were not unlike those of the major oil companies – the similarities could be striking. Ecumenical activity is wonderful in theory and yet, as so many have discovered, difficult to put into practice. There could be a shared objective but much independence of mind as to its realisation.

My arrival at these commercial cathedrals had posed a whole lot of questions that were very different to managements' usual deliberations. Instead, there was much to be thrashed out which challenged individual personal conviction and flexibility of mind.

Inevitably, the ever watchful media got wind that something 'different' was going on, and the very sort of reportage that I dreaded began to appear. It all smacked of 'soul saving'. All too often the articles were written by good and godly people specially imported to write a 'religious' piece. The oil industry was entitled to be uneasy and wonder if the prospective chaplain had a hidden agenda.

The history of the Church is littered with projects that never came to full fruition. Often their promotion required and received considerable finance, and they were directed by people who knew exactly what they wanted to do, with one fatal flaw. The initiators failed to research thoroughly the constituency they sought to serve. Instead they presumed to understand the hopes and expectation of those who were targeted. Everyone can, and does, make mistakes, but this sort of primary error can be made worse when failure is not admitted. At that point the wailing starts because there has not been the expected response to the agenda. The fact that there had been no consultation before the project was launched had been blithely ignored. This was the awful warning for me. The homework had to be done and then, at least, the Chaplaincy would be launched with the acceptance of those both on and offshore.

A great deal of time was spent in responding to management's and staff's perfectly understandable reservations. For many this was the first time (or a very long time since) a church activity had entered their lives. It raised all manner of question and revealed some of the great misunderstandings that are commonplace. Everyone needed reassurance that the proposed outreach of the Church had no complicating baggage that sought to 'win souls for Jesus'. Quite simply it had to be a mission of caring and service.

The saga continued. That progress was being made was shown by the invitations I began to receive to a range of meetings almost infinite in their variety. There could be a gathering in a large plush office with figures, unknown, draped around the room. Some of them, disconcertingly, behind me, 'Andrew, you don't mind if these blokes stay and join in the discussion?' and the unanswerable, 'I know you are most anxious to meet people.' Sometimes there would be a splendid lunch in one of the companies' private dining rooms. At least it looked like a splendid lunch, but most remained on my plate. A group of senior managers would start quizzing me, and the tasty morsel perched on my fork never reached its destination. There was no shortage of questions. 'What sort of parishes have you had?' 'Do you know anything about the sea?' (My time in the Navy was very useful.) 'What would you expect to do once you are offshore?' 'Will you hold services?' Then there were the more discursive, 'What is wrong with the Church?' 'What is right with the Church?' 'Is the Church dying?' 'What if no-one on a platform goes to church?'

I don't believe the interrogations were intended to discomfort me. Questions were voiced because answers were sought. Most interestingly, my questioners felt able to ask them because they were in a situation with which they were familiar and did not feel threatened. The company executives were playing a home match and the representative of the Church was in a minority of one. Such a situation was all too rare, and every opportunity was taken to make the most of it.

This sort of exercise could be exhausting. Not because I felt on trial, although in some way this was probably the case, but because the quality of my response was important. It was a humbling thought but I realised what I said, and how I said it, was the only way these men could come to some sort of decision.

In a strange way I began to relish these encounters that would stretch well beyond the usual lunch hour. They were a very necessary preamble before any decision about the Chaplaincy could be made, and they were a forewarning of the sort of discussions that would take place offshore.

But what emerged with alarming clarity were the misconceptions and misunderstandings about institutional religion that were mind-blowing and were a thought-provoking testimony to the inability of the Church to effectively communicate with the wider world.

The topics to be most frequently raised in the years ahead were not unexpected but very wide ranging: payment of the clergy; demands of church membership; the purpose of institutional worship; the mystery of baptism; and churches with doors closed throughout the week. All subjects that are almost taken for granted by those within the ecclesiastical system, but are a complete mystery to those who are not.

During my brief spell as industrial chaplain in Inverclyde I was involved with those working in completely unionised heavy industry. Years before, when the Church had first sought to enter the yards, the proposal that there be an industrial chaplain was put to both management and the convenor of shop stewards and his colleagues. Theirs were the hands that gripped the starter's pistol. The scene offshore was very different. There was a fragmented union presence – it has never been a major feature in the life of the oil industry – but in their wisdom oil companies would discuss major proposals with their staff committees, who were drawn from those offshore.

Reactions to the proposed offshore Chaplaincy were interesting and predictable. Those with a Service background, of whom there were many, or those whose previous work experience had brought them in touch with industrial chaplaincy, did not find the possibility of a cleric in their midst to be too peculiar. As for the rest – it was a bit like 'give the lad a chance', but judgement would be reserved.

As the days went by, I found myself looking up at the choppers ploughing their way north from Dyce on the route I would come to know so well. They would follow the coastline north from Aberdeen to Peterhead then turn north east, over the prison, and then on and out over a watery waste. I still had to meet the workforce, but all the time I was picking up bits of useful knowledge from patient office staff, who, at one time, had worked offshore. It made me decide to take a more systematic approach which would help to fill some of the many gaps in my knowledge.

It was frustrating to be waiting in the wings but I had no quarrel with the time it was taking to form a view about the proposed Chaplaincy. It was a totally new concept: the rights and views of the workers had to be respected, and the harmony of platforms maintained. The cost, not inconsiderable, had to be determined and agreed. All this within a pot-pourri of companies – each with its own distinctive ethos and style.

The 'Industry, oil, offshore' section of Aberdeen's public library came to know me well. The publications surprised me for their glaring omissions. There were loads of compelling statistics detailing how a burgeoning Aberdeen Harbour, which in 1969 had recorded two hundred and fifty movements of supply vessels, had, within a decade, increased it to four thousand. Books and leaflets informed me that 40 per cent of the gas used in the UK was piped ashore at the St Fergus Terminal. I remembered, to my shame, how I had seen all the publicity when there was the transfer to North Sea gas, without sparing a thought for the way it had come about and the immense human labour that had made it all possible. I read about Flotta in Orkney and Sullom Voe in the Shetland Isles and the massive throughput that made the oil available. I was starting from scratch but bit by bit was beginning to grasp something of the complexity of the offshore oil operation. It was a colossus and, as with most people, the immense technical challenge and mind-blowing costs had passed me by. Over the years the industry had been spending between five and six billion pounds every year: it was just as though a Channel Tunnel was being constructed every eighteen months and this telephone number expenditure reflected the intensity of the activity, the cost of maintaining existing projects and the development of new ones.

Yet behind all this investment there was an army of people. Three hundred thousand supported the industry and forty-eight thousand were directly employed by it. I found the size of the offshore enterprise to be mind-boggling, and yet as I read the literature there was little mention of the human beings who brought the statistics to life. People were 'out there' – a bland comment in the *Financial Times* of the day that 'oil production in the North Sea has risen by two million barrels a day' could conceal so much human activity. Here was a huge, economically indispensable workforce, yet to all intents and purposes it was invisible. In the light of all that was being achieved, that was unfortunate, but even worse was the lack of public recognition of those whose efforts cascaded down into its vast dependent industries.

My interest was fully aroused and I started going on courses which would fill in some of the blanks in my knowledge. Three days were spent on the theory and practice of drilling. It was an eye-opener – to see the film of the wellhead in Iran where lengths of drilling pipe were spewed out of the ground like spaghetti. It was a vivid reminder of the raw power that was ever present in oil extraction. All the processes had an inherent explosive energy and it was very obvious that everyone who worked offshore had to know precisely their area of responsibility and discharge it safely. There was no room for passengers.

In my particular case the industry's hesitation in confirming the establishment of the Chaplaincy became even more understandable in a situation where bed space was at a premium, operating logistics complex, and where critical technical problems could suddenly arise.

Although the chaplain was a minister in the Church of Scotland, for operational purposes that fact was quite irrelevant and inter-denominational goodwill was crucial.

Positive meetings were held with representatives of the Episcopal and Roman Catholic Churches, and real understanding was established, but it was not always easy and was not helped by the acute shortage of clergy which made the establishment of a joint Chaplaincy a non-starter.

At that time, Mario Conti was the Roman Catholic Bishop in Aberdeen and was personally a great help and encouragement to me. He was the quintessential churchman, and a useful ally at times when my lack of skill at ecclesiastical chess would be painfully exposed. It was no surprise when he was elevated and became Archbishop of Glasgow.

The Lord Provosts of Aberdeen (there were two in my time) were both welcoming and supportive of the many projects that came to be initiated. Their role was unusual for they found themselves presiding over two very distinct Aberdeens. There was the traditional, with its ancient university, beautiful parks (having won the City in Bloom competition for nine successive years, it was asked to refrain from entering again!) and well ordered civic life. The advent of oil wrought big changes. New businesses and a totally new industry arrived and the hitherto tightly-knit city had to loosen up. There was an International College for children from the United States, and the Dutch and French had their own schools. An Oil Club became an important location for formal entertaining and there were many places established for more informal socialising at every level. Yet, the two communities did not relate to each other a great deal. The vast new offices were there for all to see but the oil industry, in human terms, did not impinge a great deal on the life of the born and bred Aberdonian.

The oil fraternity does not put down roots. It expects to be mobile and after a few years in any one particular location it moves elsewhere – probably to the other side of the globe. The difference was yet more accentuated by a workforce which was by no means local. Most of those who worked offshore preferred to live elsewhere in the UK, or beyond.

The oil companies played their part in the cultural life of the city, sponsoring concerts, exhibitions and projects which could develop their involvement within the community. But a dispersed workforce, designated heliports (unknown and unvisited by most Aberdonians), and platforms far over the horizon, made the integration of such a significant industry, with all its spin-offs, surprisingly inconclusive. But always the Lord Provost of the day made it very clear to me that his door was always open if support was needed. In the days ahead that was to prove important.

Through energetic networking and brain-picking, I acquired a useful jumble of knowledge and I was left in no doubt that the Chaplaincy could fill a small, but important, hole in the heart of the industry. Then came the long-awaited news. I was given permission to visit the North Cormorant Platform, but this first trip offshore was to take on a complexion that could not possibly have been foreseen, because a few days before it was to take place, in November 1986, a Chinook helicopter returning fully laden from the Brent Field had crashed into the sea within sight of Sumburgh airport. Of the forty-seven on board, only two had survived.

It was not the start anyone could have anticipated. It seemed quite extraordinary that the first major offshore tragedy in the UK sector of the North Sea should coincide with the formal beginning of the industry's Chaplaincy. It was in a climate of shock and sorrow that the first offshore trip began, and the long journey to the far north gave me ample opportunity to mull over many things.

2

A Time to Remember

LOG EXCERPTS

⌕ *Long talk with the medic. He raised the question, what is the most important investment of an oil company – the machinery or the human machine, and is the same care given to the latter?*

⌕ *After this visit am fully convinced that my task is to be with the offshore workforce and to try and responsibly represent their best interests when on the beach.*

⌕ *Got home early at 1315 hours to a rapturous welcome from two dogs hugely delighted at the prospect of a wet and windy walk.*

⌕ *Arrived on board at 1330 after a wearisome flight, full of glitches. But unusually, and thoughtfully, a hot lunch was waiting.*

'Work is the most important thing a man can have.'

William Naismith Wylie
– Jarrow 1936

The orange-suited, hooded, men emerged from the warmth of the airport terminal into the wind and rain and walked in single file to the waiting helicopter. The arc lights blazed down on an otherwise deserted tarmac. I was the last in line and waited until everyone had found their favourite seat; my fellow passengers were frequent flyers for whom the whole experience was routine, in contrast to myself for whom all things were new.

The journey had started in Aberdeen with a fixed-wing flight to Sumburgh airport in the Shetland Isles where we had entered a cavernous concourse, constructed in rather optimistic anticipation of the activity to be generated by the offshore oil industry. The flight had been delayed and the building echoed with emptiness with its unmanned check-in desks and departure gates. In addition a brooding silence seemed to have enveloped everyone and everything. This was understandable. Only twenty-four hours previously a twin rotored Chinook, packed with men returning home after a fortnight offshore, had crashed into the sea within sight of the airport, killing all but two.

There was much to concentrate the mind as we all struggled into our survival suits and went through the boarding and escape routines. Rapt attention was paid to the familiar safety instructions; for everyone the purpose of the survival training had become very real, and there was the recognition that the unlikely could become reality.

My passage on this particular flight had been agreed some days before after long and exhaustive discussions. The appointment of a chaplain to the offshore oil industry was a new idea and not everyone was persuaded it was a good one. Now with this tragedy it had suddenly taken on a totally unforeseen dimension. With a multiple tragedy the industry was entering uncharted territory: emergency plans had been established and incident procedures developed, but nothing had been tried in a real-life situation. And what could not be determined was the impact of the disaster, not merely on friends and colleagues on the Brent Charlie and Delta Platforms, but on the offshore workforce as a whole.

We clambered into the chopper, a trusty Sikorsky 61, recognised as the workhorse of the North Sea. We tied our life-jackets round our waists (the safety vest had not yet arrived), put on the earmuffs that cut out the noise of the rotors but also cut us off from any communication with everyone (earphones with direct contact from the pilot had not yet been introduced), clipped ourselves into our seats and settled down cocooned in our isolation for what would be a long flight.

Winter comes early in these northern latitudes and night comes quickly with much flying carried out in darkness. There was nothing to see, nothing to hear and the old hands took out their paperbacks, composed themselves for slumber, or just thought their own thoughts.

The headwinds were quite strong and the Sikorsky, not the most aero-dynamic of aircraft, did not find it easy to slice its way forward. There was lots of time, just what I needed, to ponder the strange path that had led me to a seat in a chopper flying from the north of Scotland over an inky black, stormy sea dressed in a survival outfit that made me feel like a cross between a Michelin man and a Teletubby.

I found myself going over my life and its strange path where powerful, and usually totally unexpected events, had prompted sharp changes of direction. I had never tried to dissect my past existence quite like this before, and the longer I pondered the more it seemed to me that for all the setbacks, and mountains to be climbed, there had been a pattern and purpose to it all which only now was beginning to unfold.

My father had been an ambitious ship-owner. Note the 'had been' for he, like so many, fell victim to the economic slump of the early 1930s. My own memories of those days are not too clear, although for some reason I can remember London omnibuses with hard rubber tyres, but I do realise that I was privileged and had begun to lead the Christopher Robin style of life lovingly recorded by A. A. Milne in *Now we are Six*. I do recall a childhood with a nursemaid propelling me round London's Kensington Gardens and, I'm sure, taking me to the Changing of the Guard at Buckingham Palace. My father's business collapse was total. He lost everything he possessed and his Clyde-built ships were put into receivership. Yet in the midst of all this failure he was fortunate. The Gourock Ropework Company, who had fitted out his vessels, offered him employment as a clerk in their office on the quayside in North Shields and he was saved from the ignominy of the dole. The move to the north of England produced a revolution in the family life style with a new home and new friends but in hindsight it probably drew my parents closer together and fortified them for the difficulties which lay ahead.

The effects of the recession were massive on Tyneside in the 1930s. Coal and steel were the basis of the local economy and there was little demand for them. Pits were idle and shipyards silent. It was the time of the Jarrow march when a few hundred despairing men aroused the awareness of a whole nation as they took the road to London to convince their political masters of their plight. At that time people lived in closely-knit communities and were united in a common longing – that they be given the chance to earn an honest wage. My father took me across the Tyne to Jarrow to give me an important lesson. I must have been eight years old, yet the memory of the grass-filled cobbled streets and the gaunt listless figures huddled at every street corner is still with me. My father, a man who rarely showed emotion, suddenly said, with a tremble in his voice, 'Mark it well – work is the most important thing a man can have.' It's the sort of remark that remains with one for all one's life, and it's only, decades later, that I have come to realise how that scene of physical and spiritual dereliction, and those words have so influenced my life and thought. But we moved on, leaving the dereliction of Tyneside for the equally tragic situation on the banks of the Clyde.

The return to my father's native Glasgow was not a complete success. It had been made attractive by the lure of a better job with British Ropes. But for my parents there were all the social difficulties presented to a man who, in his own eyes, had failed in London. He was diffident in renewing friendships that had been established when more affluent times had determined a life style he could now no longer afford. Most of his contemporaries, who had, like him, survived the 1914–18 War, were now well established in the industrial and business life of the city. Understandably he was happiest with his own company. It was only when he was dying, and heavily sedated, that he relived the awfulness of his time in the trenches as a company commander in the Cameron Highlanders. Now, in 1938, twenty years later, having endeavoured so much, and worked so hard, he found himself in yet another hell that was not of his making: I wonder if the impact of the economic depression on those who were already affected by that carnage and, with great effort, tried to cope with normality, has been really understood? The days of recognised post traumatic stress disorder were a long way off. Somehow, my father managed to secure a place for me at his old school, Glasgow Academy. This may have helped to shore up his very damaged self-esteem but did not solve the problems that began to emerge.

To secure orders in the shipping world one had to be involved in quite a lot of entertaining. My father's capacity for alcohol was strictly limited and nothing like that of his prospective customers, so there followed some very unhappy years. The War had begun and blacked-out streets posed problems for the sober, and many more to those who had drunk too much and were trying to find their way home. The anxious hours with my mother – a Yorkshire woman brought up, and convent educated, in Belgium – as we waited for the sounds of my father trying to find an elusive keyhole, and then the inevitable recriminations, affected my school work. I imagine that, in this enlightened age, teachers would be alert to poor academic performance by someone who was not a complete fool. In the 1940s problems at home were 'no go' areas and my dismal class record produced despairing comments in school reports: 'This boy needs prodding and appears incapable of prodding himself,' or, rather more prophetically, 'It is to be hoped that this boy is a late developer.'

Although I did not lack friends, in some ways I had a lonely childhood, for there was much that gave me food for thought but yet I felt unable to share. But all was not lost! As well as rugby, I acquired a love of reading, not least biography. I found it reassuring to immerse myself in the lives of people, both known and unknown. In these lives there were frequent accounts of family crises of one sort or another; of weaknesses and fallibilities, and it helped me to develop a sense of proportion about my own particular problems.

As the fascinating complexities of the human spirit started to unravel it became clear to me how the whole of life is marked by its frailty. In one way or another every one of us is imperfect and even amongst the brightest and the kindest there can be some darkness lurking somewhere. I did not see this not very startling discovery as depressing, but as being realistic. From then on I ceased to be too upset by the occasional hiccup in my own, or other people's, life's journey. This realistic approach was to give me confidence much later on in my dealings with the powerful and successful. At times when discussions were quite heavy and experts were trying to exert the full force of their own not inconsiderable personalities on some issue, I found myself pondering on what the particular chink in their own armour might be. Time and again I was to discover that the strongest cement welding humanity together was not its assumed strengths but its acknowledged weaknesses.

Religion did not play a great part in my life. I went dutifully to church with my parents but struck up a deal with them that I would keep quiet during the sermon if I was excused attending Sunday School. This was a more influential decision than any of us realised for as I sat, peacefully, in the pew I would turn the pages of a Bible and found to my delight that the Gospels were not describing perfect people, but instead told an intriguing story about the Man of Nazareth who lived with, and made followers of, the most ordinary, frail and fallible, individuals. This made sense to me.

Growing up in wartime Britain was disconcerting. At school, as soon as war was declared, the senior boys set to digging up the tennis courts and constructing air-raid shelters, and elderly school-masters emerged from well-earned retirement to replace younger colleagues who had joined the Forces. For some of us it was not long before the reality of war was brought home. The school had organised a berry-picking camp in Fife in the summer of 1940 and one day a Whitley bomber, clearly in difficulties (aircraft recognition was every schoolboy's speciality), flew in from the North Sea and tried to land at Leuchars Airfield. The plane was on fire and the schoolboys in the raspberry drills watched open-mouthed as it tried to make the runway – it failed, and suddenly there was a massive explosion: a ball of fire, and then a huge black column of smoke. The silence among the watching boys was unforgettable; people could die and the horror of conflict had dawned.

In the following March, Clydebank was blitzed. Although some miles away, Glasgow's west end suffered from random bombing. A rogue landmine, described by my father as 'a parachutist' drifted by as he stood, unwisely, in the bay window giving a running commentary on all that was going on. There was a massive explosion and the glass fell in on him as our neighbouring block of flats was reduced to rubble with considerable loss of life. All I remember is a tin-hatted policeman walking into the hall through the gap that had once been the front door. We lost some ceilings and most windows and experienced the vagaries of blast, with my mattress being impaled by a shard of glass and my Cadet Corps uniform, laid out for the next day's parade, being blown into the backyard. The windows were quickly boarded up, but the shortage of glass meant we spent the summer in a house with no natural light.

The war years were difficult for everyone but memory can be merciful and erase unpleasant events. My recollections were of fire-watching at school, ration books and a stream of troops from overseas: notably Poles, Belgians and French (my bilingual mother was in her element), all spending their leave in our house. There was the occasional bitter news that a pupil remembered as a 'big' boy had been killed in action. One began to feel as a youngster that one was being 'schooled' and better able to cope with the reality of death and the indiscriminate way in which it could suddenly touch anyone's life.

There was no family tradition of naval service but for some reason I had always wanted to join the Navy. On my seventeenth birthday I volunteered, and after interview, was accepted into the 'Y' Scheme. This was designed to produce the leadership material that would man the landing craft heading the invasion of mainland Europe. However, one month after I volunteered, D-Day was launched and the need for intrepid, albeit ignorant, midshipmen was no longer a prime concern. A few months later, in November 1944, I was summoned to report to HMS *Royal Arthur* at Skegness, known to thousands pre-war as the original Butlin's Holiday Camp. There the news was broken that the 'Y' Scheme had been disbanded and the powers that be had little need of me. I was offered a choice. I could return home and enlist in the Army or I could continue in the Senior Service in one of the few categories which, at that particular time, were short of recruits. I became a Probationary Stores

Assistant. This was a shock to my system for I had been brought up to believe I should be a leader, but I had made my choice and had to buckle down and get on with it.

Most of the next three years were spent in the Mediterranean. My naval career was marked by its lack of distinction best summed up by a discharge certificate which said it all: 'This rating is clean (!) and of a pleasing personality – would have done well had he remained in the Service.' But this time was not totally wasted; it made me realise my education was far from complete. For a middle-class youngster to be plunged into a wholly different environment was, to say the least, very thought-provoking.

The unquestioning acceptance of a conventional Presbyterian, somewhat inflexible, view of life that had seemed to matter so much in my upbringing was now subjected to close examination. More usefully, I think I became less of a prig; I was slower to judge and became much more sanguine about the life styles of my shipmates. Of course, language was cruder and somewhat monosyllabic and when some matelots went ashore the nautical tradition of womanising and carousing was upheld, but without malice. There was lots of humour and minimal gossip from those who opted for an alternative way of passing their spare time. There was a refreshing absence of self-righteous comment.

In the immediate post-war months Italy was in turmoil. Civil order was not yet fully re-established and corruption was rife amongst a population desperate for food and clothing. One day I was in the back of a lorry sitting on a pile of naval stores – nothing could be left unguarded – following the ancient Via Appia from Naples to Rome. On either side stretched vast areas cordoned off with barbed wire. On the one hand was a gigantic treatment centre for servicemen who had contracted venereal disease. On the other an equally large encampment for men under sentence for black market offences. It seemed symbolic of a collapsed society where sex and the quest for a fast buck had become the norm. It was quite a shock for a just-turned-eighteen year-old and I found I was forced to reconcile my idealism, that had never been put to the test, with the reality that the veneer of civilisation could so easily be stripped away. Frail, vulnerable man stood revealed. My experience was commonplace to my generation but it did make me review the traditional attitudes that had been inculcated into me. I had to revise a lot of my ideas and sort out a lot of the hypocrisies and pretentiousness that, it seemed to me, could be so destructive of right relationships. It slowly dawned that unqualified friendship, loyalty, sacrifice, generosity of mind and humour were far more important than much of the superficial behaviour that marked conventional middle-class morality where 'being found out' could be the greatest sin, and keeping up appearances the greatest concern. Of course there could be rogues; equally there could be loveable rogues. Indeed, the more I saw of life, the more it was a trait I began to identify amongst the first disciples. Then came the turning point in my life.

The flimsy piece of paper was marked 'VAM011353 Unclassified' (I still have the copy). It invited volunteers of all ranks to apply to go on a Moral Leadership Course being held by Army Chaplains at their centre in Assisi, starting on 23 November 1945. I volunteered and much to my surprise was given permission to attend. But obtaining permission to take part and actually getting there were two very different matters, and I had eight days to sort things out.

Stores Assistants in the Royal Navy had no clout, so getting off the island of Malta where I was now based and reaching the centre of a very disorganised Italy needed a bit of ingenuity.

The journey began with a flight in a Dakota, piloted by a demob-happy member of the American Air Force, who insisted on flying round and round Vesuvius for 'one last look': The few passengers on board, who had all hitched a lift, were inclined to get increasingly depressed by his expressed wish. Once landed, a series of lifts in Army jeeps and lorries took me north through a war-ravaged country full of people in need of the basics for survival. The return journey from Assisi was to be even more disconcerting. I found myself completely disowned by the Senior Service who could not cope with a naval rating who had suddenly appeared out of the Apennines seeking passage from Naples to Malta. Eventually, I managed to get on board the Italian cruiser *Guiseppi Garibaldi*, ferrying UK troops to North Africa. It took three weeks to get back to Malta and my disappearance had been 'noted'. Apparently, I had given every sign of being a Christmas deserter.

Far more important than the journeys was the three-week course in the home town of St Francis. The team of six Army Chaplains were very special. They had served in the North African and Italian campaigns. They were highly experienced soldiers with a rich denominational mix, and they were united in one goal; translating what could be a confusing Christian faith into something intelligible and relevant. I was ready for their talks and devoured their reading lists. C. S. Lewis was the writer of the moment and I wrestled with *The Problem of Pain, Beyond Personality* and *The Screwtape Letters*. Has there ever been a more effective apologist for the Christian faith? Equally rewarding for me was to mix with thirty soldiers of all ranks who had seen, and done, so much. I was the youngster in their midst. Indeed a shopkeeper in Assisi thought that, in my uniform, I was the first British schoolboy to visit the town since peace was declared! British naval uniform in Central Tuscany was somewhat unusual. Talking for hours with these men was to learn from a group whose maturity extended far beyond their years. The young sergeant in the Scots Guards who couldn't wait to return to run his pub *The Cheshire Cheese* convinced me of a hunger for a spiritual dimension in facing the challenge of daily living. It all made such sense to me and I found myself wondering what part I could play in reconciling the world of the Spirit with the world of work. It dawned on me, relentlessly but without any inner fireworks, that I should seek ordination in the Church of Scotland – the one denomination I knew.

There were compensations during the years in the Mediterranean. For a time I worked alongside Geoffrey Shaw from Edinburgh, whose naval career was not dissimilar to my own. He gave my thinking an edge. Apart from a memorable encounter in a Naval boxing ring – volunteers can do the most inexplicable things – we had many equally rigorous meetings – verbal ones! We would spend hours developing our hopes and visions, for we shared the same spiritual convictions. Geoffrey had a first-class mind and clarity of thought which was, eventually to lead him, thanks to a Commonwealth Fellowship, to the storefront churches of Harlem in New York City, and then to a return to a one room flat in Glasgow's Gorbals with an ever-open door. Eventually, he became Convenor of Strathclyde Region and Scotland's foremost political leader. His lamentable death in his early forties deprived Scotland of one of its greatest sons. For myself, I continued to hasten slowly, having put together a rather more prosaic agenda.

If you wanted a place at university, not unreasonably some proof that you could read and write was required. My erratic academic progress at school meant I left without this vital evidence. The 'open sesame', the University Certificate of Fitness, was denied me until I had passed some formal exams. Each time I returned to Valetta in Malta and with the encouragement of the resident Naval padre, Hugh Purves, who was to be a great friend and mentor for decades, I would sit a subject. Miracle of miracles, by the time my demob number came up I had accumulated enough points and I was able to apply to Glasgow University. I spent the next six years there in a mixture of frenetic study and blissful freedom. I still treasure the moment when my former headmaster crossed a busy Glasgow street to hail me and say that the prize I had been awarded had 'given the school a particular pleasure but, I have to say, a certain degree of surprise'.

In the immediate post-war years universities were filled with a wide age range of students. There were those straight from school; the real ex-service veterans who had endured six years of war and the 'not quite' veterans like myself, who served longer than the National Servicemen but whose actual war service was minimal. The compound of age groups and experience produced varied attitudes to student life. Those who had just left school tended to keep their heads down and immerse themselves in their studies. Those who had left school years before simply wanted to get on with life and cast off their student days as quickly as possible. I think it was my age group who gained most from university. The three years, or thereabouts, in the Forces were not dissimilar to an extended gap year. New experiences, if little real danger, eased the change from schoolboy to student. Now, the requirements – wide reading and extended essays, no matter the subject – were quite different and over six years the combination of Arts and Theology constituted a formidable challenge. I found it a mind-stretching and demanding intellectual discipline but, most importantly, I was taught how to think, which surely should be the basis of all university education. Social life is best described as 'intense', with the vacations being spent as a bus courier with Scotia Tours, conducting six-day trips round my native land. The Wake's Week clientele from Lancashire were the most demanding from the moment the ladies donned their paper hats bearing the interesting information 'I am a Virgin Islander' (quite racy for those days).

The most significant personal discovery was that demobilisation euphoria had concealed the reality that I still had to do a lot of growing up. Even so, all the necessary exams were disposed of, and I found myself on the clerical market.

3

Getting Involved

LOG EXCERPTS

✎ *It suddenly struck me that the one thing you miss as you go round the accommodation module, are the personal touches in the cabins, i.e. family photographs. The average worker looks on his platform as a place of work where he happens to be trapped and therefore has to sleep there.*

✎ *Few folk had the slightest idea who the Moderator might be. One man asked him if he went to church.*

✎ *A man spoke to me about his broken home and his relief that he was offshore at Christmas. The prospect of not being able to see his children would have been unbearable.*

✎ *Platforms are rather like time capsules and the events that govern our lives; Christmas, Easter, bank holidays, Autumn breaks have no place.*

✎ *So much for my dieting! The man behind the counter where you draw survival suits had one glace at me and said 'XL'.*

'Life's the only teacher.'
Offshore incident survivor

The early 1950s were boom years for the Church. There were thriving congregations and a healthy support for all its traditional activities, social and spiritual. In June 1953 I was ordained and began my ministry in Whitehill parish church in Stepps, an ideal place for a young man to start his ministry. It was sufficiently flexible to let him find his feet and express himself, yet resilient enough to survive his wilder ideas. In the next six years I learned much, not least in being privileged to share many of the crises that suddenly can strike any home. In those first years I was ill-prepared for the continuous involvement with death and its accompanying grief. The process of dying, and the shock and sense of loss endured by those left could be awe-inspiring, as were the countless examples of quiet courage and nobility of soul. These situations were not easy to cope with when one really cared for all involved, but gradually I came to accept that we are all in a terminal state from the moment we are born.

Presbyterian services can be marked by an absence of congregational participation, and I found this to be unsatisfying. It emphasised a personality-centred approach to worship that made me uncomfortable. Even at this early stage, I found that services seemed to be at their most meaningful when I was at sea as a chaplain in the Clyde Division of the RNVR. Everything appeared so much more natural and worshippers valued the opportunity to be with God in surroundings in which they felt at ease. There were times when it was almost a physical effort to return to traditional ministerial life with its seemingly endless cycle of funerals, marriages and baptisms. The hours spent at wedding receptions, chomping my way through tomato soup, steak pie and sherry trifle, often seemed, in spite of all the networking, time that could have been better spent. To my shame I discovered how cynical I could become about the most significant events in peoples lives – the hatches, matches and despatches that tied them to the Church by a cultural, rather than a spiritual, rope and explained why it was often the source of unreasonable expectations.

I found it hard to work out what the role of institutional religion was supposed to be. My first apprehension of God, years before, as I was rowing from the Isle of Tanera to Achiltibuie, on the shore of Wester Ross, seemed far removed from the interminable meetings devoted to the maintenance of church buildings. Despite everything that vexed and confused me, the sense of the reality of God never deserted me. I found that to hold the hand of someone who was dying and had already glimpsed the glory that is everlasting life, never ceased to humble me, and confirmed, to me, that spiritual realities are the only ones that really matter in our ephemeral existence. I may have been confused about the institutional Church, but not about the reality of God's love for all when, through his Son, he inserted himself into the stream of human history.

I remain ever grateful to a congregation who were sufficiently resilient to survive my questioning and who helped to establish within me a belief in the unique relationship that is Christian fellowship. This was to stand me in good stead in the years ahead. But after six years it was evident that everyone needed a change – there is a limit to congregational flexibility! Suddenly life took one of many unexpected turns – perplexing at the time – but afterwards making a lot of sense. The continental connection raised its head.

My mother had grown up in Belgium where her Yorkshire father had suddenly decamped after a 'vision' where he rebelled against selling pots and pans from a horse and trap in the Yorkshire Dales and, after reading about the marble-cutting skills in Belgium, decided to set up a factory at Alost. He must have had a very benevolent bank manager for he had little money, no linguistic ability and knew nothing about marble. But his vision saw that material as being fundamental to interior building and design in the UK. Astonishingly, he was proved right and he went on to fit out, amongst much else, the bathrooms in the Savoy Hotel and the Tearooms of Joe Lyons.

My mother came from Methodist stock, but residence in Belgium resulted in an education by Irish nuns in a local convent. Her side of the family continued to live and work in the country for many years (with breaks for two world wars). The congregation of The Scots' Kirk in Brussels were looking for a minister and because of the family connection I thought I might fit the bill. I didn't! There had been a revolution in Iraq and the chaplain to the oil companies suddenly found his parish had disintegrated. In the more politically turbulent parts of the world such events were not unknown, and it was freely accepted that any minister unavoidably affected had first choice of vacant overseas charges. Alistair McLeod went to Brussels and I stayed in Stepps, still mildly intrigued by my impulse to seek to work outside Scotland. Soon afterwards, out of the blue, I was asked if I was interested in going to either Lausanne or Lisbon, both of which had long established Scots' Kirks. Portugal meant little to me but Switzerland was a different matter; set in the heart of a Western Europe that had just negotiated the Treaty of Rome. It was evident that there was going to be a time of convulsive economic and political change with Lausanne right in the centre of it all.

I arrived in Switzerland in June 1959 where the congregation in Lausanne turned out to be elderly and determinedly expatriate. The numbers were not large and the 'welcome social' took the form of a cocktail party in the not over-large apartment of the session clerk. However, the English-speaking community was about to be transformed. Large international companies had begun establishing their European bases in the city, and through the benevolence of the Nestlé Foundation, IMEDE – the International Business School – was established, attracting the finest minds to create an outstanding teaching faculty and drawing the brightest and best young men (very few women in those days) from international industry and commerce who would spend a year in concentrated study before returning to their sponsor organisations. Many faculty members and students came to worship on Sundays and created a continuous challenge in presenting the gospel within the context of business, society and the individual.

Being resident in Lausanne in the 1960s was very special. It was the era of John F. Kennedy's 'new frontier' – best epitomised when Gisela, my German secretary, came into the church office one day in floods of tears having just heard the President's broadcast from the Brandenburg Gate in Berlin, where he announced in ringing tones 'Ich bin ein Berliner.' And it was a time of contradictions. The Cold War was at its height yet the United States sought to develop its markets in Western Europe. With all its business tax advantages, Lausanne was an attractive place for companies to base their operations and many household names set up their marketing headquarters manned by high-flying staff. It was not long before the church was bursting at the seams with bright young men and their equally bright wives and children. It was necessary to start up an English-speaking primary school – now the International School – and at the same time there was a large influx of students from all over the world who wanted to learn French. With all that was going on, my learning curve took off vertically.

Lausanne was an international city, beautifully situated, facing south across Lac Léman (the Lake of Geneva) to France. It had attracted English-speaking residents for decades. Their arrival could cause endless speculation, for there could be a multitude of reasons apart from study and the strategic thinking of multinational companies. It was possible for individuals to conclude ingenious income tax arrangements with the authorities which could be financially advantageous. Also, there must have been something about the Alpine air which attracted men and women, notably from the USA, whose personal lives were in crisis and sought some relief from their traumas. In stark contrast there were those unfortunate expatriates who had retired to Switzerland when the pound sterling was strong, who had no Social Security, and who saw their small pensions hugely reduced with the decline in the exchange rate and now lived in very real poverty.

Superficially the expatriate community gave an impression of abundance and stability, but lurking below the surface was much doubt and uncertainty which could suddenly emerge, prompted by an alien culture and unfamiliar language. Britain and the United States must be amongst the most linguistically inept countries in the world. Both countries seemed to believe that if you could not be understood it was merely a matter of shouting louder. This rarely worked and could prove counter-productive, with mutual irritation and a frustration that could abort the best of intentions.

It was fascinating to glimpse something of the different methodology of large organisations and the pressures upon their employees to perform in exchange for generous compensation. In many instances, however, the high spending power produced its own difficulties, as couples sought to cope with an enhanced life style (an anticipation of the North Sea 'golden handcuffs'?). Pastoral problems could be dramatic and I developed an arrangement with the local gendarmerie who would come and collect me – invariably in the middle of the night – if 'une situation' looked as though pastoral and linguistic backup might be useful.

Living far from home in a sophisticated international community could strain relationships. People became emotionally vulnerable and could do stupid things. To my surprise I found myself a regular visitor to the local prison, and on occasion to Bochuz, the high security penitentiary that served the Cantons of Vaud and Geneva.

The congregation was a potpourri of nations and denominations and many thought-provoking questions were raised which demanded a response. For most of the religious traditions present, the celebration of Holy Communion was the focal point of regular worship in their home country and they could be mystified by the comparative infrequency of its celebration in the Church of Scotland. This was such a contrast to the regularity of an act that was the bedrock of their personal religious experience. I saw all too clearly that not all spiritual truth can be verbalised and it is the combination of word with action that means much to so many.

Equally interesting was the challenge to religious communication, and the many questions it prompted. Why was it possible to take sixty young men and women for a weekend hike in the Alps and discover how moved they could all be by an act of worship in a mountain refuge? Was it the Orders of Service with their responses that I had humped up the slopes? Or maybe it was the uniqueness of the setting? Or was it the discovery that saying one's prayers could be found to be entirely natural? Again, why should twenty extremely busy men, who had been all over Europe during the week, be prepared to spend a couple of hours once a month on a Friday evening, reflecting together on the problems, business and personal, and often international, that they faced in their journeying? Why were these same people happy to attend weekend seminars on ethical decision-making? Maybe in all these things there was a novelty factor, and being away from one's own country can do strange things. But, more pertinently, in everything there appeared to be an acknowledged relevancy, and the meeting of a need. I realised that within the Church one could be constructing agendas for its faithful and then assume there would be acceptance. But where was the listening and the meeting of expressed needs? Could it be that without being aware of it I had become more selective than I realised and was out of touch with views other than those of the most supportive, and usually, uncritical members? All the time I was learning. Take one example: I had to radically reassess my understanding of Remembrance Day, which had been mildly jingoistic until I noticed from the pulpit two girls, not known to each other, sitting together in a pew. One had lost her father, shot down in a bombing raid on Hamburg. Beside her was a German girl whose parents had perished in a firestorm in that same city.

The international make-up of the expatriate community was a broadening experience, as was the denominational mix within the congregation. Rubbing shoulders with people you liked but who represented different religious slants, or none, put everything into perspective. It was not a question of being bland about the differences; neither did it mean that one had become a woolly liberal. It was simply an acceptance of the reality of a world where God spoke to his children in so many different ways. It was clear that if people were to live in peace there had to be a recognition, and acceptance, of ethnic and religious differences. The supreme ignorance was to harbour a racial arrogance and treasure religious bigotry. In a fast evolving world there could be no accommodation for such views.

The eight expatriate years were a badly needed broadening experience but whilst it answered some of my questions, it posed just as many. Theoretically it seemed to make some sense in 1967 when I was appointed the first General Secretary of the Scottish Churches Council, based in Edinburgh. My contacts with so many nations and denominations would argue that I was the man for the job. Theory and reality can be so different, and in retrospect I cannot think of anyone less suited for involvement in the labyrinthine deliberations of Inter-Church Committees and Ecclesiastical Commissions. Committee work has never been my forte. This was my fault because I have a total absence of the necessary conviction to do a good job in such situations. I had become a firm believer in the evolutionary way things would work out in the Church Universal, but my new responsibilities involved me in the more forceful way of inflicting agendas on the faithful who had not been educated in the thinking of the ecclesiastical world and, understandably, were not convinced of the need for change. Conferences of the convinced, the ecumaniacs, were all too frequently held, and then with great publicity dates would be announced for the Union of Denominations. Time would pass, nothing would happen, and then these dates would be conveniently forgotten. All this time a watching world looked on with increasing incredulity. It was inevitable that the understanding between those outside and those within the institutional Church increased from a gap to an abyss.

Church politics can be a minefield, and I was forever getting blown up! It required an ingenuity of mind and a patience I did not possess. One day I overheard a church dignitary remarking that he found 'Wylie to be a bit too bland!' Of course he was right. So often I felt the public voicing of opinions and attitudes to be a complete waste of time. My interest and concern was for those who got on with their response to a challenge with the minimum of fuss and publicity. It seemed to me the Church was big on announcing what it intended to do, but small in initiating what it sought to achieve.

So much of the ecumenical posturing was unintelligible to most people whose personal faith was uncomplicated and unquestioningly assumed within the perspective of a religious tradition which had been determined by accident of birth. This simplistic view will not be acceptable to some, but my own experience suggests that it is the emergence of apparently dissociated and peripheral events (hunger, poverty, disaster, common goals, passionate concerns) that bring people together in strong bonds of unity. A friend told me of the time his church was celebrating an important anniversary and he invited an eminent divine to preach the sermon. 'How did it go?' I asked. 'Very well,' was the reply, 'the congregation didn't understand a word, but were quite happy to let him wrestle on their behalf with something that was obviously very important to him.'

My time with the Scottish Churches Council was not a happy one, and it suffered from my lack of leadership. But the consequences for me were not as serious as they might have been, for an occurrence in my own life, which was never to be the same again, exerted an incidental but valuable damage control.

Having a stroke is a bewildering experience. I was walking up The Mound in the centre of Edinburgh on my way to the Scottish Churches Council Office when it happened. There was no pain, but an instinctive realisation that something enormously traumatic was going on. I overheard a passerby saying to his companion that 'life must be difficult for spastics' and I realised that my left leg was dragging and my left hand and arm were totally useless. Somehow I hung on to my briefcase and struggled on to my place of work. Unlocking the door presented problems: it would make a good party game for a man to extract his keys from his left-hand trouser pocket when only allowed to use his right hand.

The summer of 1970 in Edinburgh's Royal Infirmary was salutary. I had always enjoyed robust health and to be lying in bed at the receiving end of all manner of ministrations was humbling. Also, it was an emotional time, with lots of tears and personal anguish as I tried to work out why this calamity had happened to me; how I was going to provide for a wife and four daughters; and how not to become a burden to others.

Being paralysed down one side was inconvenient, but I was fortunate, unlike my visitors, because I had not lost my power of speech and I was able to write. Concentration was a bit of a problem, but over the long months ahead I had plenty of time to reflect on what life might yet have in store for me. I never felt inclined to shake my fist at God, but curiously, I found myself asking why the Faith to which I was so committed could make people so insensitive.

In hospital, helpless in my bed, I was prayed over a great deal by the representatives of many denominations. Belonging to an inter-church organisation had some unforeseen implications. These caring clerics provided a procession through the ward (these were the days of twenty-bed wards) which was a source of endless interest and speculation to my fellow patients. On one occasion a visitor, with an exceptionally loud voice, boomed, 'Andrew, let us pray,' and I noticed that, as one man, the rest of the ward pulled their bedclothes over their heads. Conversely, a minister I hardly knew, other than the fact that life had not been easy for him, said a few words, obviously coming from the depth of his own personal suffering, and I felt greatly strengthened. My worst moment was the afternoon when I had been dozing and awoke to see a figure clad in black standing at the foot of the bed holding a black notebook. For one awful moment I wondered if this was what the day of judgement could be like, then a voice said, 'Oh, sorry, you're not on my list.'

Little things came to matter a great deal. One morning I woke up to discover that the second hand of my self-winding watch, which had been inert since The Mound incident, was moving. Great excitement! It meant that I had been moving my arm in my sleep. This was the start of the long road to full recovery.

There were glorious days when I would be wheeled out on the ward balcony to take in great gulps of fresh air. It was August, the sun shone, and the green of the trees in Edinburgh's Meadows was like a benison. My confidence began to return, and there was another blessing. It was the time of the BBC Promenade Concerts and every evening, isolated by my headphones, I would let the sounds of Brahms, Beethoven and my beloved William Walton wash over me.

It was a time of cleansing. I was reminded of the importance of simple things: the beauty of God's world; the healing through human compassion; the reality of a loving Father, and the resilience of the human spirit if hope is not surrendered and the will remains positive. Bit by bit my mobility returned and constant exercising with a rubber ball made sure that, eventually, I developed a formidable left hand grip. There was a long way to go, but I felt a much more complete individual – not least at being able to distinguish between busyness and productivity. The whole convalescence period was very good for my humility.

When my father was a ship-owner in London I had been baptised in St Columba's Pont Street. However, never, ever had I imagined that I would return there as a temporary replacement for a legendary deaconess. It was Fraser McLuskey, the minister, who early in 1971 invited me to take her place and whose sensitive understanding gave me the opportunity to rehabilitate myself. I lived in a compact little flat in the church tower in the heart of Belgravia and bit by bit was able to play some part in the ongoing life of an unusual kirk with its gathered congregation drawn from all over the metropolis, and beyond.

St Columba's was a magnet to the streetwise Scottish homeless. No doubt some were in great need, but many had spent years getting to know 'the system' and milked it for every benefit they could get. In the basement of the church there was an impressive array of very wearable clothing donated by a kindly congregation. One day I took a client downstairs to the shower room, and took away his verminous old garments to be burned. I left him happily soaping himself in the hot shower with the promise that I would return and completely kit him out. Two hours later an agitated visitor came in to the church office and announced that a naked man was in the basement clinging to a radiator to keep warm. It was an active church and very easy to get diverted from the business in hand!

However, I really excelled myself when sent to collect a cast-off suit belonging to a distinguished member of the Court of Appeal. A house-keeper directed me to his Lordship's bedroom saying the suit was ready for collection. The garment was lying on the bed and I swept it under my arm and departed. The phones were red hot by the time I returned to St Columba's. In my enthusiasm I had removed the eminent judge's best suit which had been laid out in preparation for an evening function – I had ignored a rather more modest brown paper parcel on a chair.

As ever, there were a whole heap of impressions to absorb. Talking with the homeless put into perspective my own troubles and anxieties, and brought home to me as never before how extreme the change of direction could be in lives that, at one time, must have seemed so solid and secure.

Whether it was the aftermath of my illness or an innate cussedness I am not too sure, but I was finding it increasingly difficult to contemplate going back to the life and work of a parish. That wonderfully evocative Scots word 'scunnered', best described my view of the Church. There was an excuse for my sourness. After I came out of hospital it came as a shock to discover that when I was unable to work there was no support system available from my own denomination, in spite of my being appointed to a recognised ecumenical body. I was married with four children of school age and as I queued up at the local labour exchange ('We rarely seem to get people in your category, Mr Wylie'), I experienced the humiliation and sense of failure that is endured by so many who are made redundant. A visit to Church Headquarters to the appropriate department had not been helpful. 'I'm sorry you don't fall into any of the categories we can assist,' was not good to hear and it angered me when I had to meet with ill-trained people who were unable to cope with the authority that came with their particular remit.

I found myself wrestling with the dilemma of whether a return to traditional ministry merely meant that, in a very subtle way, I would be contributing further to the perpetuation of a system that my experience told me was dangerously cut off and afflicted with a myopic sense of its own importance.

I went to see the then Chairman of BP, Bobby Fraser, and suggested to him that his vast Britannic House in the centre of the City of London was one of the new cathedrals and that it should have a chaplain. Although I was obviously angry and confused, I was not totally laughed out of court, even though there were practical problems that made such an appointment impossible, but it was arranged that I meet with the Director of Human Resources. This was a crucial meeting for me. After a long discussion he said he could offer me a job, and then came the interesting bit, 'but I think you would be miserable because you would miss the very special relationship that, as a minister, you enjoy with people'. That very prescient remark halted me in my tracks and made me realise I was wallowing in self-pity, and in my discomfort with the institutional Church I was completely over-looking the priceless relationship between clergy and people that has been established down the generations.

I had to take a deep breath and start again, accepting the springboard that was already in existence and from which much could be developed. There was a delicate balancing act that had to be established between working within the traditional patterns of ecclesiastical life and witness, and at the same time making a meaningful contribution to the world of work in which all God's children were involved.

The direction to be taken suddenly became clear when in the spring of 1972, quite out of the blue, a church in the heart of Edinburgh made noises suggesting they were looking for some new thinking to inject meaningful regeneration into their historic pile.

St Andrew's and St George's was a union of the first two churches to be built in Edinburgh's New Town. It had fallen on difficult times with its doors firmly closed during the week while the world of work passed by. These same doors were opened for a few hours on the Sabbath onto a more tranquil world, where most of the pedestrian traffic was generated by the insatiable consumerism that can be the mark of Presbyterians going to the place of worship of their choice.

Once installed in the church, I realised that things were not going to be easy when a minister, not known to me, crossed a busy George Street, and without any preamble questioned my sanity because 'city churches have no future'. This might have been true if one was prepared to accept that institutional religion was only practised by the middle classes, on Sundays, for the most part tucked away in their suburban fastnesses. That was not how I understood the Incarnation, with its clear assertion, that God came in his Son for the good and delight of all men and women. It seemed to me that the great challenge was to show that God is for Monday and that he is for everyone.

Bernard Camp, who was a Jew, was the Manager of C&A on Princes Street and he summed up the aims of St Andrew's and St George's very succinctly one day. At one of the many 'getting to know you' meetings he remarked, 'You have great site potential – develop it!' Over the next decade we tried to do just that. Excavating under the church enabled us to create an under-croft with a comfortable lounge area and cafeteria and, most importantly, a chapel, where office workers who came to enjoy a light lunch had the opportunity to join in worship together, or sit by themselves in quiet con-templation. Its very presence was a testimony to the naturalness of prayer and devotion in the midst of everyday life.

It took time and perseverance but eventually most of the major stores in the centre of Edinburgh came to accept the idea of a store's chaplaincy. The gestation period was long and not dissimilar to the start up of the Oil Chap-laincy, and the hesitations and caution were a foretaste of the North Sea adventure. In the earliest days I found it took a bit of readjustment to stand in the corner, of say, Ladies' Coats, and discuss some important personal issue with a member of staff. I did not find it easy, but I discovered there was no sense of strangeness from employees – the feeling of awkwardness was entirely mine. I drew much comfort from St Paul when I remembered how he had written about being a fool for Christ's sake.

The change of direction 'hastened slowly' and there was a tangible recovery of a sense of purpose from a loyal and committed congrega-tion where a high proportion became involved in one project or another. For myself the biggest problem was inevitably being associated with 'the big church' in George Street. It was a privilege to have this link, but the downside was an association that presented difficulties in understanding and acceptance, for the perception of the Church was, for the most part, astonishingly negative and uncomprehending. Yet . . . there was a gleam of encouragement. A, so-called, secular society was happily prepared to support a representative of the Church who sought only one thing – to serve a community where it worked.

My marriage had, over the years, been running out of steam. Not with angry words and bitter exchanges, but through an almost imperceptible, yet relentless, slide into two lives that moved along in co-existence and it finally ended. By this time, three of my daughters were grown up and had left home and the fourth was nearing the end of her secondary education. After fourteen years it was time to leave St Andrew's and St George's. This was confirmed by a splendid old lady who remarked, 'Oh Mr Wylie, at last we will be able to get a real minister.' Then seeing my expression she patted my hand and added, 'Mind you, you were very good at funerals.'

I became the industrial chaplain for Inverclyde – a task for which I was entirely unsuited (for the second time!). It required a political astuteness that was outwith my experience and I was the despair of my long-serving colleagues in the industrial mission team. In my simplistic view, it seemed that if the quest for social justice was to be meaningful and durable, there had to be a wellspring of deep Christian concern which was quite separate from any particular political or social standpoint. I learned much from splendid men at every level of responsibility, not least the convenor of shop stewards at Lithgow's Yard who was gifted with a huge breadth of vision, and the oratorical gifts to do it justice.

Scotland can be very special in the men and women it produces who have no educational advantages and yet have powers of insight and intellect, together with a love of learning for its own sake that can make them so singular. I was reminded of my session clerk in Stepps: John Lithgow, a railway signalman for forty years and possessed of the finest natural intelligence I have ever met. It was reinforced by a robust theology which he had developed through reading and his personal devotions. Typically, he readily came with me in the mid 1950s to a meeting that sought to encourage industrial mission in Glasgow. We were the only two in the audience.

Can it be that the Church has failed to identify this reservoir waiting to be tapped? It is sad that so much that is uttered from the pulpit fails to ignite the minds of its hearers, let alone their hearts, and 'the hungry folk look up and are not fed!' As I was to discover in the North Sea, too little attention was paid to the innate spiritual depths dormant within so many. Thus an opportunity for mutual enrichment is lost.

I had only been a few months in Inverclyde when the shaping on the anvil called life was to receive yet one more nudge. It began in Edinburgh at a committee meeting. This, attendance at which for me was a rare experience, began, like so many before it, with predictable enthusiasts making their predictable contributions. All of a sudden I began to pay attention, for the talk had moved to the North Sea and the lack of contact between the Church and the offshore worker. Great play was made about the responsibility of a National Church for those who nominally fell within its ambit. It dawned on me that the possibility of ministry in a situation devoid of all the usual ecclesiastical trappings was very appealing. I found myself putting the case for the establishment of an offshore Oil Chaplaincy, and before I sat down I heard myself accepting the suggestion that I go to Aberdeen and test the ground to see if an idea could be translated into reality. Within committees there can be a considerable lapse of time between the launching of a concept and its implementation – but if someone offers to kick the ball into the park, a lot of enthusiasm can be generated. In a remarkably short period of time I was seconded from Inverclyde to Aberdeen . . . and so it all began. Several years later, I married the fair-haired Jennifer, who became known in the North Sea as 'wee blondie' and there developed a sustaining partnership through some of the offshore oil industry's most emotionally demanding years. She was to join me on her first, and my last, trip offshore.

* * * * *

My personal soliloquy came to an abrupt end as I peered through the cabin window of the helicopter. Giant flares seemed to be reaching up to the low clouds and everything was bathed in the intense reflected light, and not so far below a lumpy sea glistened like wet coal. Dominating the astonishing scene were the vast, brilliantly lit platforms, standing proud on their steel legs.

For my travelling companions the wonder and the awe it prompted had long since gone. But to a new arrival the Shetland Basin was like a city of the future – H. G. Wells fulfilled. Down below, thousands of men and women spent half of every year of their working life.

There were lights everywhere. Those surrounding the helideck made it look like a gigantic fried egg, and the floodlights that brought the drilling rig into sharp relief made me think of the Thames Embankment and an illuminated Cleopatra's Needle. Lights blazed from large, anonymous, cube-like structures rather like compressed Christmas trees.

As we flew in, the helipad disappeared from sight directly beneath us. My mind went back to all the meetings that had led to this moment. I was acutely aware that the Church had no divine right to go offshore – the privilege had to be earned. There was a gentle bump, the whine of the rotors ceased, in the relative peace we filed out onto a wind-blasted helideck. My first steps were very tentative, not with the fear of being blown over the side, but with nervousness about what I had embarked on. In a very small way, my fevered imagination made me feel like the astronaut who took that famous 'first step for man' as he landed on the moon.

4

Those on Board

LOG EXCERPTS

⌕ *The OIM is an Israeli and quite a polymath with linguistic, artistic and culinary gifts. He was quite clear that being a 'people' person was far more important for an OIM than a high level of technical expertise. The proof of the pudding . . . his own crew were tangibly well adjusted.*

⌕ *There was a red alert during the morning because of a suspected fire in one of the legs. Helicopters were put on the alert for a possible evacuation. It was interesting to observe the behaviour of everyone at the muster point. These gatherings are now very much for real and I personally find them to be quite demanding – one thing for sure, the clergy are expected to behave in a selfless way!*

⌕ *An hour in the Wendy House. Good talk and much leg-pulling of the 'walking on the water' variety. 'Andy, when you come to see us why do you bother with a chopper?'*

⌕ *There were few women on Fulmar. Those who were stewardesses were felt to improve the atmosphere and to be a real asset. There were two young girl apprentices – not so mature and probably less able to sustain balanced relationships. It will be interesting to learn how they work out.*

⌕ *Was called this morning at 0445 with a glass of orange juice. A compensation for being kept up so late.*

'The Chaplain shall be the friend of all on board.'
Queen's Regulations and Admiralty Instructions

☞ *Talked with galley staff. Much concern, as they see it, about the inequity of the terms and conditions of employment and the need for a core figure who could organise labour. There is no question that there is a deep feeling that the workforce is being exploited. I heard the phrase 'tartan coolie' for the first time.*

☞ *Just what do you do when there are 120 people on board and the one recreation area accommodates 18 people?*

☞ *Some offshore folk become 'fat cats' being used to good wages and large chunks of time off. There does seem to be a corrosion of values, and some men have become so spoiled that basic gratitude and a sense of fulfilment have been replaced by a feeling verging on 'divine right', that the Company owes them something all the time. Some of the expectations and complaints were quite ludicrous and provided an 'awful warning' of what can happen to people when their sense of values goes into hibernation.*

☞ *Had an excellent luncheon in the FSU and much talk with everyone. Discovered that apart from the foreman of the deck crew (from Stornoway), I was the only other person who could splice a rope – times have changed – nowadays hawsers are clamped.*

'☞ *The value of the medics should not be underestimated. Even when they are noticeably, and happily, underemployed. They still are the one reference point within the crew who is 'with' but not 'of' them.*

☞ *The driller is an obese man with a particularly blatant collection of 'pin-ups'. I don't think it's his chauvinism that offends me, but he must be the first man I've met offshore who, quite simply, I would not trust.*

The popular image of the offshore worker is wide of the mark. The public imagination sees him as a bulky individual, a bit of a boozer, and with a touch of the Rab C. Nesbitt about him. The sort of person who might have a problem holding down a job on the beach. There is a loudmouthed beer-swilling minority who do a disservice to their colleagues, although there have been times when my respect and affection for the offshore worker has been less than enthusiastic. As I began to get used to the offshore life I came to understand the sense of release after a fortnight of that unlikely mixture of total predictability and hidden, unspoken, tension. Whenever I was taking a south-bound train from Aberdeen, I tried to hide behind a newspaper in case a stentorian voice cried, 'Hi there Andy, come and join us,' and I would find myself at a table submerged in six-packs. It's unfortunate that such a bibulous time could be one of those rare occasions when the general public were able to identify the offshore workforce both visually and it has to be said, aurally. On such occasions their misconceptions would be reinforced. Inevitably there are misfits and undesirables but eventually the disciplines of the life and their colleagues suss them out and they return permanently to the beach.

Although there is no such creature as a typical offshore worker, there are certain idiosyncrasies that identify them. Travellers on planes and trains might be aware of the casually dressed passenger beside them – invariably more reflective – on the trip north to Aberdeen. He is unlikely to initiate a conversation, but will willingly join in one, if invited, for they are people with extremely forthright views. The astute observer will notice that often there is that inner stillness that hints at a disciplined mind. It was always interesting to watch these men disembarking from commercial flights. Instinctively they would slow down the unseemly rush towards the exits by standing in the aisle and letting those in front go first. A working life where boarding and disembarking from helicopters is an integral ingredient can inculcate habits where calmness and good order get the best results, not least in an emergency.

Physically, the offshore worker comes in every shape and size, from bulky human mountains to the lean and wiry. I was constantly amazed at the unflagging energy of the slim and slightly built who kept going at the most strenuous jobs on the drilling floor, just as the very bulky could show unsuspected agility when climbing into a packed cabin on a Bell helicopter for a brief inter-platform transfer. The differences were not only in physique. The range of accents was spectacular. Sing-song voices from the Welsh valleys mingled with Glasgow's glottal stop, and the special vocabulary of Buchan was juxtaposed with the inimitable lilt of Geordie land. Britain's rivers were well represented, Tyne and Tees flowed along with the Mersey, the Clyde and the Tay and one could sense the ghosts of Britain's long gone industrial greatness. The oil industry owed a great debt to a redoubtable industrial heritage.

Quite a few of those who work offshore pursue learning for its own sake. A not inconsiderable number hold Open University degrees, and the reading material to while away the time on the tedious helicopter flights was remarkable. The best of written material was avidly consumed – one Bear was working his way through the Waverley Novels with great determination. Knowledge of current affairs was astonishing – usually being given an annual airing at the Platform Christmas Quiz.

The life offshore is attractive to those who are not afraid of their own company. How many other people spend four hours each fortnight, year after year, physically cocooned in an all-in-one suit and head-phoned – deaf to all but an occasional message from the pilot?

The oil industry was very conscious of its macho image. Probably it was a transatlantic importation and one that some of those who first worked offshore sought to reflect. This rough, tough, hard-drinking, womanising creature of fiction was emulated and invariably attracted media attention if there was trouble on the beach. But as time has passed a whole series of events have tried and tested everyone; sometimes as individuals and on other occasions en masse. It has often been shown that courage and determination have nothing to do with physical size and strength. I had a particular respect for a pint-sized steward who could not swim and yet had passed the survival course which included an underwater escape from a submerged mock-up of an inverted helicopter cabin. He was always cheerful, unruffled and pleasantly unhurried. His influence for good was palpable and I am sure I was not alone in reckoning he was the man to be with in a time of crisis.

The pendulum factor is ever present in history. Attitudes and events never remain static but are in a constant state of flux from one extreme to the other. The human pendulum has noticeably swung in the oil industry and nowadays I detect an almost anti-macho culture. It is a sign that the industry has come of age and those who work in it do not have to prove anything. A quiet confidence has replaced the earlier brashness.

All the time there were the surprises. One day on my return from offshore, I was wriggling out of my survival suit at Dyce heliport and went through an eye-popping experience. I watched a crew member emerge from his protective overall resplendent in an immaculate suit. From his holdall he retrieved a felt hat (crush proof) and strode off to catch an onward flight, looking for all the world as though he had just negotiated some big business deal in an Aberdeen boardroom.

Over the years the number of women working offshore has steadily increased. Stewardesses and galley assistants were in the van and were an interesting lot, widely travelled. Many of those who maintained the living quarters had worked on cruise ships and cargo liners; others had been employed by international catering firms who thought nothing of organising a birthday party for thousands in the Moroccan desert, feeding troops in the Falklands, or running a residential compound in Saudi Arabia. As the years have passed the work opportunities have now widened. On the larger platforms visitors can be greeted by a receptionist who, amongst much else, allocates cabins and hands out keys in the best hotel tradition. Female engineers are in charge of shifts which could include a girl apprentice. Some of the women in admin divide their time between offshore and onshore offices. The unwell can be treated by a female medic and the hale and hearty will have the best the chefs can provide. But being a woman offshore was not always easy going. One girl, a telecom engineer, explained how strange it had been to sit in splendid isolation in an otherwise crowded cinema. The men felt they would leave themselves open, at best, to leg-pulling, and at worst gossip, if they sat beside the girl. Happily, these days are now long past and there is a unanimous view that the addition of women to the crew has done nothing but good. There are all the predictable jokes about an increase in the sale of aftershave. It's probably true, but it all goes towards making a more presentable and fragrant (!) crew, when off duty and OIMs were united in their view that a mixed crew did much to raise living standards.

The Norwegian platforms have had a strong female presence on their platforms for many years, and a Saturday evening dance was not unknown on the Statfjord Platforms, which were in sight of the Brent Field. A young steward on his first trip offshore was steadily dripped information about a big dance that they had been invited to attend. With the exception of this young lad, whose enthusiasm for 'the dance' grew with every hour, everyone else on the platform was party to the great deception and when the loudspeakers announced that those who were going to the dance should go to reception, everyone watched with interest as a freshly shaven, shoe polished, trouser pressed, clean shirted individual presented himself, enquiring about the departure time of the helicopter that would take him in five minutes to the North Sea Palais de Danse. After a few hours he realised there would be no dancing that night – or any other.

The female influence could express itself in the most unexpected ways. One day I was doing my rounds on a platform and was shown, with great pride, the latest piece of machinery that was to be installed by the female engineer and her team. It was a massive new safety valve; the sort that gladdens the heart of the Health and Safety Executive. As my guide burbled on he began to induce in me that all too familiar numbing sensation that descended whenever I was submerged by a mass of technical detail, but I was preserved from total coma by the colour of the new piece of machinery. To say it stood out is an understatement. Colour schemes in the North Sea are garish and somewhat lacking in subtlety. In the midst of all the vulgar colour it sat, coyly confident. It was painted a most delicate pink, best described as Elizabeth Arden's 'Rose' – a wonderfully subtle feminist statement.

Those who worked offshore could be as reticent about themselves as French Foreign Legionnaires. Many appeared to make a deliberate attempt to compartmentalise their lives and keep work and home completely separate. The result was that men could work together and share the same leisure time and yet not know one another as individuals. One day I was talking with two men to discover that one was a trained cabinet-maker and the other made furniture as a hobby. They had served on the same platform for several years but had never discovered their shared interest. The coincidence didn't stop there. They found that their fathers had been Japanese prisoners of war; even more remarkably they had been in the same camp. This bridge building became possible once it was established that asking questions arose out of a genuine interest and not a prying curiosity.

Spontaneous group discussions were refreshingly revealing. One evening the subjects ranged from Scotland's football prospects to the nature of religious belief. All verbal passion spent, the group, eventually, reluctantly dispersed. The OIM, who had been an enthusiastic participant, remarked, 'Fascinating, I've learned more about these men and women tonight than being with them all these years.' No doubt the group were saying the same thing about their boss.

Sometimes a crew would seem like a gathering of the United Nations. One Christmas Day the dinner was prepared by a Malaysian chef, and my shared cabin was looked after by a steward from Chile. When I visited another platform I discovered the OIM had been a major in the Israeli Army. Linguistic ability could be tested. It seemed incongruous to be in the middle of the North Sea and to be discussing the Scottish Rugby XV with two French-speaking engineers from Senegal. Indelibly etched in my memory was an impressive driller from Canada – a native Canadian – who brought with him something of the stillness of his forebears. He was a quietly devout man and I drew much strength from him.

All this internationalism was taken for granted – oil men think globally – and knowledge of countries and continents meant that news items were often given a different slant. Far from the popular concept of a world of rough, tough, boorish men, it was one of lateral thinking and broad horizons.

In their own communities a wide range of community involvement was undertaken. One evening offshore I had a fascinating meal with a group of workers who included a local magistrate, a captain in the Army Cadet Force, a yacht builder and the captain of a long established cricket club. These sorts of interests and responsibilities were not unusual and were an indication of independently inclined, well motivated, resourceful people.

For many the two weeks at home were more than occupied. There was the usual domestic routine to slip into, where wives went to work and children to school. There were elderly parents to support, house painting and gardening to be done, prize-winning vegetables to be cultivated, pigeons to be raced and dogs to be bred, football teams to be supported, school functions to be attended and, most importantly, anniversaries to be remembered and celebrated. A great deal could be crammed into the fortnight on the beach.

As I wandered round the North Sea's platforms it seemed as though urban man had become maritime man. However, there had not been a total parting of the ways from a work culture that could be traced back through the generations and where, significantly, worker representation was accepted as a fact of life. When disputes did arise, and considering everything they were remarkably few, I did wonder if the oil industry had ever taken the time to understand its workforce, which was quite different to any it might have employed in other parts of the world.

Unsurprisingly the Navy and the Royal Marines were well represented offshore, being particularly experienced in dealing with helideck and flight despatching duties. Old habits die hard, and the give-away for many an ex-matelot was the rather smart light- weight overalls that were sported by them when off duty in the accommodation unit. But the extent to which those on a platform became welded together, like a good ship's company, could vary enormously.

The fortnightly crew change brought this into sharp relief when the whole ethos of a platform could be transformed. This was partly due to the arrival of new personalities but primarily was the result of the handover to the newly arrived OIM whose management style could be vastly different yet just as effective. It must be one of the more remarkable features of contemporary management when the leadership changes on a relentlessly regular basis yet the hassles are few. The flexibility and sensitivity required to make this possible says much for the character of all concerned.

Being involved with work situations for a long time makes one very aware of the pulse of an organisation. There were times when one noticed how the emphasis on technical competence was progressed at the cost of the development of people skills. The quest for increased productivity and cost efficiency could result in a blurring of the focus on those who worked at the sharp end and their leaders. Every airport bookstall sells a plethora of volumes on sales and management techniques. A closer look always shows that the space devoted to books with a human, non-exploitative, emphasis seemed to be minimal – maybe not much is written about people for people? For the most part the companies I was involved with were exceptionally 'people orientated'. The fact that they were 'happy' companies was no coincidence.

Alarms are the norm for most platforms, not least those extracting oil and gas in high pressure fields. I began to realise that the appearance of relaxation could be superficial. Wariness and alertness were rarely discussed but were traits that ran through everyone's minds. The capacity to live in a constant state of readiness depended on the individual. It had nothing to do with being macho, but everything to do with realism and professional confidence.

If there is a crisis – great or small – on a static installation, in the middle of an empty ocean, the loneliness of decision-making comes home. To be having a light-hearted conversation with an OIM and then watch his face if the standby alarm sounded, was to watch a man transformed. The features showed awareness that the next few moments might prove to be the ultimate test. Then shortly afterwards the 'stand down' would be given and the same individual would revert to the man you had been talking to.

The OIMs were an interesting lot. Every company had its own method of selecting men for those key appointments, and it resulted in an astonishing array of experience, skills and natural talent. There were Merchant Navy skippers with a tanker background who were used to transporting combustible cargos and had 'safety' engraved on their hearts. They were probably the most at ease with their appointments. The Senior Service was represented by retired naval officers who were very conscious of the need for a 'happy' ship and the cleanliness that went with that desirable state. Maybe this insistence on 'bright work' was not always understood in a world where industrial discipline had to be exercised with considerable diplomacy for there were no Queen's Regulations and Admiralty Instructions to back up action against the wayward and lazy.

Invariably there were those with an engineering background whose career paths demanded they spend some time offshore. Often they seemed to find some of the demands of management to be trying; not least where human issues were involved. Engineers were great pragmatists and often it seemed that 'soft' issues did not come naturally.

Within this eclectic assembly there were two outstanding groups of impressive individuals. The first came from comparatively humble beginnings in the oil industry and had climbed steadily up the responsibility ladder. En route they had acquired vast experience of men and machines. The second group were the high flyers, a smaller group of OIMs whom the Almighty appeared to have endowed with an unfair dollop of talent and who spent a few years offshore adding 'OIM' to an already comprehensive and impressive CV.

No matter the route taken, once the appointment was made the OIM assumed an awesome responsibility. In an instant he could be called on to make decisions that could affect lives. Events can develop with great speed and the complete isolation of an installation can make heavy demands on personal integrity and professional competence. There are several hundred of these very special men (as I write women are just starting to push through this particular glass ceiling) who are at the sharp end of the UK's oil industry and they deserve much more public recognition.

To walk round a platform with an OIM could be revealing. Sometimes everyone was known by name and all the work in hand was understood. Others liked a tidy platform free from all litter (could it be rather like Captain's rounds?). Some OIMs valued meal times as opportunities to get alongside their crew and they would make a point of eating at the busiest time and sitting wherever there was a spare seat. For others, meal times were precious moments of peace and quiet in a hectic day and valued as times of solitude in a very people-crowded life. It was all a matter of style and personality. There were those who sported a collar and tie; witness to a sartorial elegance, ignored by others.

It was no nine-to-five job. A platform manager could spend many hours a day at his desk – fifteen is probably a conservative estimate. Phone calls from the beach would be arriving all the time on every conceivable subject, and there was a constant flow of people seeking work permit authorisation, which before it could be issued required a detailed explanation of what had to be done and why it was necessary, and there could be times when complex technical discussions ensued before the necessary permission was granted. Always there was more than the work in hand.

It was not unusual for the OIM to have to break bad news. Death and accident are sadly part of the ingredients of living, and when misfortune took place it was the OIM who headed the complex arrangements that included transfer to the beach for the unfortunate crew member, and securing a suitably qualified replacement. Disputes had to be dealt with immediately. If they could not be resolved then the ultimate disciplinary sanction had to be invoked and a return to the beach arranged. Much of an OIM's life was the sort of man management that is familiar to many, but always there was that additional 'North Sea Dimension' where safety could never be compromised. OIMs were a realistic lot who could be relied upon to react decisively in any situation that might arise. As a group of individuals they were fascinating in their diversity, some giving very little away about themselves. That was the 'other life' and kept quite separate. Revealingly, not all of their desks bore reminders of home and their families.

Each OIM had a very personal interpretation of their roles, which were as different as their personalities were distinct. For all the idiosyncrasies, the astonishing reality was that the wide varieties of leadership all worked. Not least, when I talked with these men, I was always conscious that in the event of a cataclysmic incident, their responsibilities meant that their chances of personal survival were less than for anyone else on board.

The medics played a key role, although they could be seriously underworked for long periods, for at any moment a life-or-death situation could develop amongst the crew. They belonged to a great tradition of service, which must be one of the rich veins in the history of offshore operations. When I visited the Piper Alpha Platform I used to be held spellbound with the stories told by the medic who had worked offshore since the earliest days of the North Sea saga – the days when men went offshore in suits and their city shoes. A survivor of the Piper Alpha incident told me that the last sight he had of this same medic was of him clutching his first aid bag and running towards the heart of the inferno. Just what you would expect – which may be easy to write but so selfless to fulfil.

'They also serve who only stand and wait' could best describe a medic's role. Everything could be transformed in an instant and he could be the most needed man on a platform. But standing and waiting did not suit some cost-conscious companies who wanted all their employees obviously employed at all times. There was a decision that the medic should be involved in some admin duties. Superficially this made a lot of sense but there was a downside. Once the medic was involved in the making up of rosters and allocating flights to the beach he became sucked into the system, and an invaluable independence was surrendered. It's the old story of having to decide how important it is to have someone on board whose prime value is just 'being there'.

Industry finds this a difficult calculation to make, and one that can only be addressed if the claim that 'people are a prime concern' is a strongly held belief and not just a glib phrase in a company's annual report. 'Waiting' has its uses and it was significant that many a worker who had no physical ailment found the privacy of the sick bay – where the door was closed as a matter of course – to be the one place where he could unburden himself. A North Sea platform is not a private place, and the excuse of going to see the medic could cover a whole host of needs. The medic was perceived to be his own man and it was to everyone's advantage that he remained so.

The medics were an interesting group of experienced nurses who had the calmness and detachment associated with their profession. They were shrewd but invariably kind and understanding in their assessments. In background, they covered the whole spectrum of their profession from nursing in the armed forces to work in psychiatric and geriatric wards. They had all seen a lot and were very conscious of the debit side of their offshore role, which had taken them out of the mainstream of their profession. To ameliorate this many would spend their time on the beach keeping up to date with the latest medical advances, and they would take the opportunity to keep their hand in at general nursing.

They were very fortunate for they had the most spacious accommodation on board, with a single cabin next to the sick bay which was well equipped with all the most up-to-date resuscitation equipment. Trips offshore could be suffocatingly uneventful with the most common call being the removal of foreign bodies from crew's eyes. But, conversely, there are many people around today who can vouch for a critical intervention that saved a life.

My first encounter offshore with a doctor was unexpected, and somehow it fitted that he should be wearing a black eye patch. This piratical figure presided over a small hospital on the North Cormorant Platform that lay some distance to the north east of the Shetland Isles. Robert Stephen was a man of many talents and he was constructive in his times of waiting. I am the happy possessor of two of his books in Scots verse, 'By the Shores of Galilee' (stories from the Gospels) and 'The Fables of Aesop'. When this particular offshore appointment was terminated, the North Sea lost a character who, allied to his professional abilities, had a huge interest in the richness that life offers and brought a much valued humanity and compassion to his duties.

Doctors are now flown out from Aberdeen to any emergency. The trip might involve a touch-down on a helipad and in exceptional circumstances, a descent by winch. Exceptional men with a remarkable response time and rather taken for granted.

People concerns on a platform came within a triangle, at the top of which was the OIM bearing ultimate responsibility for everyone and everything. But there were two excellent base supports in the form of the medic and the

Camp Boss. This latter landlubber title was transported offshore and harks back to the times when catering companies first began to meet the needs of oil workers in desert compounds. There is a vast difference from those early days. The convoy of supply lorries has been transformed into steel containers that are loaded on to the supply ships and eventually hoisted onto the platforms. The containers are packed with everything imaginable (and a bit more) that can bring physical comfort and wellbeing to those offshore. Quite simply, if there were no Camp Boss there would be no containers.

Invariably they were men of great experience; 'seen it, done it', types. Usually they had worked in the most unlikely places. They had the overall responsibility for managing the catering, maintaining the living quarters, and operating the shop. The latter sold tobacco (smoking was permitted in parts of the accommodation module) and all those toilet bag and clothing extras that in a fortnight away from home can demand replacement. Apart from the sweets and underwear, it was surprising to find a very sophisticated range of scent and choice cigars. Simply explained, so far as Customs and Excise were concerned the platforms stood in a duty free zone. Cigars and scent would be purchased before crew change and sales were good.

If one stands behind a shop counter, or watches someone loading up their trolley at a supermarket checkout, one can build up quite an accurate profile of the purchaser. It is the same on a platform. Purchasing patterns, even of the most everyday items, could reveal quite a lot and if you then add the insights gleaned from responsibility for the mess room and living quarters, the Camp Boss becomes a man with a remarkably accurate picture of the diverse personalities that make up an offshore crew.

There was a kindly thoughtfulness about these men who, maybe more than anyone else, seemed to have developed a sixth sense for people who were troubled. One day I was taken aside by a Camp Boss who expressed his concern for one of the deck crew whose determination not to show a deep personal distress had succeeded with everyone apart from himself. It was a sad story of a cherished son who had suddenly died. For all the heartbreak the grieving father had not wanted to let anyone down, but the return offshore was proving difficult not least because of his concerns for his wife. The perceptive and compassionate concern of that Camp Boss made the North Sea, in this particular instance, a much warmer place.

This notable trio – OIM, medic and Camp Boss – by the nature of their roles and their very different areas of responsibility, were entirely independent. When combined with their experience, it gave them a unique perspective of life on a platform that I found very helpful.

Nowadays psychological profiling can be seen as a very necessary part of team building. I was always intrigued that, apart from the standard medical, there was a complete absence of any obvious selection process in the composition of crews who could be working, and living, together for many years. Human chemistry can be potentially combustible at any time. When you put a very diverse group of men and women together for fourteen days in a steel box that is going nowhere, it is not unreasonable to expect personality clashes, yet they were minimal. Maybe the location of the platforms produced its own filters? There were the survival and fire courses to be gone through, and the flights in the chopper. These activities were not to every aspiring offshore worker's liking. It had been known for prospective crew members to put on a survival suit for the first time and then feel unable to walk out to the waiting helicopter. There had even been times when a chopper touched down on a platform and a new arrival had one look at the helideck perched above the waves, and had decided the life was not for him, and he would remain on the plane and return to the beach.

There was a potentially combustible mixture of company operating staff and long- and short-term contract workers. Put them together and there would seem to be many opportunities for friction. But this was cancelled out by an abundance of good sense, and lots of humour, which oiled the machinery of living and working together. Despite occasional on-shore incidents that might suggest otherwise, the mark of the offshore worker is self-discipline. It has to be that way, for there simply is no room for outbursts of destructive emotion.

There was a big gap between the public perception of those who worked offshore and reality. On my very last trip offshore I was joined by a professional scriptwriter who was doing research for a TV series that was to centre on the offshore oil industry. He was very frank about his investigations and said he was particularly keen to unearth tension and aggression. His trip coincided with a farewell party for myself and I am still curious to know what he made of it. The crew gathered in the cinema and out of the blue I was presented with a water-colour of their platform, and before the presentation one of the crew played a couple of Bach fugues (very suitable for a chaplain!) on the keyboard. All very moving for me, but not the display of emotion being sought by the emissary of the BBC.

The conditions of life offshore demand that it be very controlled. There were times when I found the tranquillity to be quite unnerving. Of course, there was the unending noise from hard-working machinery, and the natural elements rarely held their peace, with the wind in ceaseless confrontation with the gas flare which marked its eruption from the bowels of the earth with an unending roar. There was more! As a platform is a metal cube, if anything of any size was dropped, everyone knew about it. But for all the racket, several hundred men could go about their business astonishingly quietly.

In spite of the confined space, it was quite possible for a crew member to spend a fortnight's trip without meeting everyone on board. Shift patterns had something to do with this. I could share a cabin with someone and yet we would never meet. There could be sounds of slumber behind the bunk curtains but I was never able to put a face to the snores! Drilling teams were particularly marked in emphasising their independence and their office tended to be a bit of a closed shop. These men still worked twelve-hour shifts; midday to midnight and so on. This was a hark-back to the days of desert oil exploration when working conditions were arranged so that both drilling crews shared the extremes of temperature. On board everyone else started work at 6 a.m. Nothing could be more effective in making sure that the two groups would never meet. However, when the alarm went off such splendid independence was shattered. Suddenly it was possible to glimpse all those on whom you might depend in an 'operation hope not' scenario.

A platform could be crammed with expertise and specialists came from all over Europe bringing a range of languages and accents; I found the West Country burr of the crane drivers to be delightfully incongruous. The specialist crane company did its recruiting in Hampshire and all points west, latching on to a reservoir of expertise that had been established over many years. To me there was a wonderful incongruity between the rigours of the North Sea and a style of speech that always evoked, for me, mental pictures of sun-dappled Hampshire meadows and hedge-lined Devon lanes. Of necessity crane drivers were physically isolated individuals sitting in their lofty eyries with splendid views of a restless sea battering away far below. For long hours they could be in their cabins seemingly perched in space. Suddenly all would be action and they would show an astonishing skill in raising and lowering loads from a great height. Anyone who pulls up the weights on a pendulum wall clock will recognise that the chains have a will of their own. To watch a crane driver control the twists and turns of a cable hoisting a steel container when the wind was strong, the swell heavy, and the deck crew on the heaving supply vessel one hundred and fifty feet below dependent on pin-point accuracy, was to see a virtuoso performance.

High up, the crane driver would peer down and judge precisely when the crane should take the strain as the vessel fell away in the swell. If he did not get it right the crane, together with his cab, could be pulled into the sea. These unloading sessions were like a piece of modern ballet. It was the combination of teamwork, artistry and technical skill that always thrilled me, and the drama was heightened if the operation was carried out on a winter's night under arc lights. No choreographer could conceive such a setting, with the foul-weather-clad handlers on the deck of the supply vessel leaping from one container to another as they hooked up the slings whilst the vessel moved in three-dimensional response to the heaving sea. The music that accompanied this balletic scene came from the constant wind and the scene had the Faustian backdrop of the flare casting huge vibrant shadows over everything.

Maybe this account of an everyday occurrence can seem exaggerated, but there is a peculiar beauty to some offshore activity. It's an amalgam of people working at full stretch allied to great technical skill, and the harnessing of elemental forces; and all of this set against a backdrop of blissful maritime calm which can change rapidly to one where the elements are at their most violent.

So many jobs offshore are hazardous. The nautical steeplejacks – the abseilers – work in situations on the platform structure that a mountaineer would describe as 'exposed'. They paint, and repair, areas that would otherwise be difficult to maintain. Their more earth-bound colleagues, the scaffolders, construct intricate structures that make it possible to work underneath the deck or over the side. Sometime the structures are like works of art, and as with all good artists there is an instinctive desire to step back and admire what you have created. This may not present much of a peril in a studio, but when you are seventy feet above sea level and you have forgotten to clip on your safety harness, it is an instinct to be discouraged.

One day a scaffolder did step back to admire his handiwork. Miraculously the stand by vessel, which had already been called in close because of the outboard work, a normal safety procedure, watched his descent (the only possible word). Equally providentially, he was immediately plucked from the waves alive. He was hoisted onto the platform twenty minutes later – thanks to the crane – and, very shocked, lay in the sick bay. He was immediately evacuated to the beach but in a few weeks had returned. He was determined to carry on, yet haunted by his failure to observe the most basic of safety rules. 'You see, Andy, at the survival school they said I was a survivor – and I am.'

Moving round the platform was the way to get some idea of the perils that went along with many tasks. I was consumed with curiosity about all that was being done and often found myself alone with a member of the crew. Much work was carried out in isolation, in complete contrast to the accommodation where privacy was rare. Isolation encourages confidences and I learned a great deal about individuals. I found it thought-provoking that the best place to relate to people was at their work. There could be a relaxed atmosphere, with the benefit of the worker getting on with his job and avoiding eye contact without embarrassment. Rather like two people travelling in a car where full-faced dialogue is out of the question and when confidences can flow so easily.

Most platforms rest on massive hollow legs. These huge, cavernous, vertical tunnels could easily swallow up a double-decker bus. Inside they are damp, dingy and dangerous. Some are fitted with lifts which save exhausting climbs up vertical ladders when crew members go to check the pumps that transfer the oil stored in huge reservoirs. The legs can be the height of St Paul's Cathedral and major problems can develop when things go wrong. Lethal gas, hydrogen sulphide, can accumulate and there have been times when workers have been overcome by the fumes. Sadly, in extracting the victims it has not always been possible to prevent fatalities. Talk to those who work offshore and their view would be that these hazards go with the job, but as the incomer I was always being humbled by the many examples of unassuming courage. Maybe this is a particular characteristic of the offshore worker. He displays the inner confidence that comes with the knowledge that when things go wrong he can depend on his colleagues. Very few hear about the perilous descent and exhausting return climb up hundreds of feet of vertical ladder by men clad in recovery gear, with oxygen cylinders strapped to their backs.

Like anywhere else in the world, North Sea operations had their fair share of grumblers and nitpickers all with a remarkable capacity to find fault with everything and yet who still persisted in going offshore. They could be endured, but with impatience. Like all experienced whiners, they were unable to give credit where credit was due. There were too, the undesirable foul-mouthed individuals who sought to create pools of unjustified unrest. They would be 'encouraged' not to return from the beach. NRB (not required back) was an edict that had to be carefully used. Happily, employment law safeguards the working rights of the individual but, equally beneficial for the maintenance of offshore harmony, a judicious use of the expulsion clause could work wonders.

Inevitably the passage of time has produced a new generation of offshore worker with no experience of traditional heavy industry and no awareness of the distinctive traditions that grew up with it. The new workforce comes from a completely different work ethos bearing no comparison to the experience of those who first worked in the North Sea and found themselves taking part in an exciting and completely unfamiliar working environment. Precedents have now been in place for a long time and much of the work is predictable. Gone is the feeling of advancing the frontiers of petroleum discovery and development. Now there is little glamour in a job which is perceived to be just another way of earning a living.

Platforms today are merely seen as factories – purposeful, efficient and powerful, devoid of personality – and the different ways of working are reflected in changes in attitude. This can be best illustrated by the crewing arrangements over the Festive seasons which traditionally were split so that everyone could have either Christmas or New Year at home. Then came the day when this was not always acceptable. On my last Christmas offshore, I was on a platform where the crew change resulted in forty-nine 'no shows'. To the old-timers this unwillingness to accept a split shift was incomprehensible. Yet, perhaps there is a cold logic in the different attitude that makes sense. It is a change in point of view and not of spirit.

A cursory glance at any platform notice-board will show that charities are well supported. Posters tell the tale of guide dogs purchased and kidney dialysis machines installed. The weekly contribution to good works is impressive and I had a first-hand example of this when I escorted HRH the Princess Royal on her first trip offshore. I had thought I knew exactly the programme that was to be followed, as did the organisers. The guided tour went according to plan, with lots of impressively well- informed questions being asked. In the mess room the informal buffet was going down, in every sense, very well. Suddenly, there was an interruption to the scheduled proceedings. A member of the crew came forward and presented a large cheque, made up from the crew's individual contributions for Save the Children. Maybe there is a new generation but it's still the same old heart.

5

Sea Cities

LOG EXCERPTS

I suspect that the 'macho' image is perpetuated by those who spend the least time offshore.

Statfjord B Platform is very spacious. All areas free of all impediment; all drilling is top drive. Legislation only allows two persons per cabin. A conference room on each level (9 in all). Games, TV, music, all in separate rooms. The chaplain has his own office (!) and spends up to a week on a platform. He has a much more integrated place in the scheme of things, taking part in training sessions and initiating schemes for more positive relationships.

Interesting talk with a refrigeration engineer from London. His first trip offshore and he found the contrast with West Drayton to be somewhat extreme. Interesting to see the industry through his eyes – quite unaware of the magnitude of the offshore operation.

When the time came to catch the shuttle, retraced my steps, over 200 of them, going down this time to the mezzanine level and then along innumerable galleries to cross over the bridge joining the Charlie Platform to the flotel. It's only on these occasions you realise their sheer immensity – the flotels are vast too but much lower in the water and you stand on a deck the size of a football pitch peering up at the vast overwhelming structure.

The admin offices on the platform are tucked away in an area quite hard to locate. One is

> *tempted to wonder how much thought, at the planning stage, goes into where the offices of key personnel are sited.*
>
> ☞ *Accessibility of those in charge, communication, and good food, are probably the main ingredients that make for a first class platform.*

When the offshore oil industry first started, the platform legs – the jacket – would be towed out to the location of the proposed development and submerged with pin-point accuracy over the wellhead. Then began the assembly of a structure on top of the jacket which, when completed, would tower above the sea. This operation was repeated many times all over the North Sea. An army of men, based on a flotel, would then take over and build the superstructure, including the accommodation module, the production plant, the drilling area and the helipad. Then would come the hook-up so that power, generated on site, could flow through the whole structure, energising everything from the control room with its banks of computer screens, to the communications systems, the laundries, the galleys, the pumps, the drilling units, the cranes and safety systems. All this was carried out in the middle of an ocean infamous for its turbulence. Given a bit of luck it would be carried out in a 'weather window' when the North Sea would be more benign.

In the early days when there were no precedents, this was particularly complex and perilous work, but as the years passed and hard-won experience was accumulated, everything became more straightforward. Enormous cranes were developed that could lift up to ten thousand tons and more complete prefabrication on shore became possible. To take one example, a complete, and completed, accommodation module could be towed out to a site and precisely installed in a matter of hours. It was rather like a gigantic game of Lego with a speed of construction that could be dramatic.

One day I was sitting in the mess room of one of the more modern platforms, admiring the sea view through a picture window. A few miles away across the ocean a platform was a scene of great activity. My coffee-drinking companion gestured towards it and remarked how things had changed during his fortnight on the beach. Whilst he was at home, the complete platform had been assembled, with one prefabricated unit being placed on top of another. They had arrived on site so complete that the installation was ready for its operational trials almost immediately. Instead of marvelling at a remarkable bit of construction, my companion was merely irritated that the sea view was not what it had been. I could have sworn that he muttered something about the North Sea becoming horribly overcrowded.

To arrive by helicopter gives no real indication of the size of a platform. Choppers tend to approach from a good height and then make a vertical descent. A mild bump can be the sign you are resting on the helideck yet all it was possible to see on the approach was a tangle of machinery and a flare which, to my over-developed imagination, seemed all too ready to engulf us if we flew too close. The arrival itself can be a bit of an anti-climax. Usually it is far too windy to admire the view and the first impression is not so much one of size but sound.

Apart from the dominating flare, there is a cacophony of clanking as pipes are moved, containers stored, and hammering from someone somewhere justifying their existence. And, always, there is the noise of the sea as it swirls around the legs.

An approach in darkness can be spectacular. If it's a clear night, from a long way off you begin to understand why platforms generate enough power to light a small town and in the blackness the flare (again!) dominates the scene, casting huge, living, shadows over the sea and superstructure. It's the combination of sight and sound that can make a platform so dramatic.

There was one time when my arrival on a platform gave me a real idea of its size. I had arrived on the flotel moored alongside and access to the platform meant climbing a series of steel stairs. My ascent began and halfway up I put down my bag, took a breather, and looked up at the massive structure towering overhead. This was the second time I'd had this view. The first was when I was on the Claymore, Piper Alpha's sister platform, two days after the disaster of 6 July 1988. On that occasion I was standing with a very traumatised OIM on the mezzanine deck just above the waves. He kept striking the huge steel structure towering above him and repeating, time after time, 'impossible'. It had always been maintained that the total obliteration of such an enormous installation was inconceivable, just as there had been a belief that HMS *Hood* was indestructible.

Platforms sit four-square on the seabed. Being multi-purpose – a place of work and rest – they exemplify everything about offshore: a curious amalgam of the small and personal and the gigantic and impersonal. They are completely immobile and cannot be taken into dry dock for repair and renovation. When this is necessary, a flotel, with a workforce of many hundreds, comes alongside to get the job completed as soon as possible. Like so much offshore, these rectangular, self-propelled craft seem larger than life. The deck area is huge and the temporary workforce, who sleep and eat on board, each morning stream across a linking bridge, which is automatically raised if the sea swell reaches a certain level. Watching them in the early morning, I could almost hear the factory hooters calling everyone to their work.

The platform and the flotel were so different: one very vertical and dependent on its height to do its job and ride out the storms; whilst the other – extremely horizontal – was built to accommodate many personnel. The latter looked most un-nautical, yet these flotels were capable of crossing the seven seas and both bore vivid testimony to human ingenuity and were eloquent reminders of this new and foreign world in which I had arrived.

The geometry of platforms is dictated by location and the meteorological rigours it will encounter. The basic structure has to stand at least one hundred feet above sea level. The 'big one' – the rogue mountainous wave – has to be anticipated and even a typical storm – Beaufort Scale 9 – could make its presence felt. I was sitting in the recreation room one noisy night doing my best to hold my own with some masters of Trivial Pursuits, and not doing very well. It was distracting to hear the waves slapping against the base of the accommodation module. I was acutely conscious of our height above sea level and the size of these seas that were making their presence felt. What made the situation bizarre, and concentration difficult, were the sauna-like conditions in the mess room. The prevailing winds had totally changed direction and the gale was directing the flare directly overhead. Everyone else was relaxed but the padre was hot and bothered!

Platforms have not the slightest resemblance to ships. They rest on the seabed, not going anywhere, and are dead in the water. A ship is vibrant with life, creaking and groaning as she faces the elements, and if the weather is notably violent she can turn and face what is coming head on. From the moment a platform leaves the construction yard it has to be towed to a precise location and, once settled in place, has to rely on its intrinsic strength to survive.

In violent weather I felt totally helpless. For me it was not easy to sit in the warmth and silence of the living quarters and remain indifferent to the turbulence outside. The tell-tale swaying of the water in the cabin WC, and the restlessness of the wind gauge in the radio room, gave an indication of the violence outside. Yet, for all the insulation, you could not forget that you were miles from land, effectively marooned on a steel island that might seem comfortingly massive, but in reality was a minute speck in a large ocean.

An excellent visual aid to an understanding of the offshore scene emerges if you fly on a clear day from Aberdeen to Stavanger. Suddenly, thousands of feet below, the eye picks out a tiny structure surrounded by empty sea. It takes quite a leap of imagination to visualise the two hundred men and women going about their daily work far below, hearing the drone of an aircraft overhead, and not sparing it a thought.

There is a chain of these steel islands that starts one hundred and sixty miles north-north-east of Shetland and stretches south to twenty miles east of Bacton in East Anglia. Geological factors have dictated the sites, and often oil and gas-bearing reservoirs have been discovered in clumps, with the Shetland Basin the most spectacular. To begin with, these platforms were operated by the household names in the oil business, and every area and each platform had its own distinctive ethos.

It is not difficult to fall in love with a ship – maybe the fact they are called 'she' is a help? I doubt if the same affection can be extended to a platform. They are notably impersonal; great big immobile beasts, never accorded any femininity, and strictly functional in purpose and design. Yet there are platforms that I could identify, even if blindfolded; it's all to do with their location.

Some of the structures are little more than pumping stations. They rest on a less complex geology and are the ocean equivalent of those donkey pumps that nod away in the Oklahoma countryside. Such platforms were always cleaner and quieter and markedly free from vibration. A considerable contrast to the installations where reservoirs were being developed and there was constant drilling and unceasing noise.

Living quarters were bedded down with rubber insulation but that was not the ultimate protection from the transmission of reverberation in an all-steel construction. Dropping a heavy spanner was the equivalent of making a public announcement! If installations operated above high pressure reservoirs, they were more prone to emergency alerts. Musters in these circumstances could be quite nervously exhausting – not least when there were a lot of them. Everyone realised that the slightest technical hiccup could set them off but one just never knew. Care and maintenance were continuous when coping with the more complex extraction systems: there would be more visits by supply vessels, and far more obvious activity and much more industrial mud and general grime. I had to remind myself constantly that with the exception of those directly involved, few people were witness to this astonishing activity.

The very first action on arrival at a platform is to attend a safety briefing when, ironically, the latest arrivals are told how to leave it in a hurry. The realists amongst the new faces immediately realise that in the unhappy event of something apocalyptic happening, no matter how comprehensive the training and self-possessed the individual, it would need a huge stroke of luck for an evacuation to be successful. The only sure way off was upwards from the helideck.

Sometimes I felt a bit depressed as I listened to the safety officer detailing the locations of the ropes that could be flying over the side to reach the swirling sea below. The depression intensified when I discovered they were known as 'the ropes of no return'.

Control of access, whilst very necessary, could be a mixed blessing. Bank inspectors can suddenly descend on a branch, HM Prison Inspectorate have the right of instant admission at any time to the institution of their choice, factory and school inspectors likewise. Not so with platforms. To get there, flight plans have to be filed and passenger lists forwarded. There can be no such thing as a surprise visit and this can send the wrong message. On some platforms there were times when I sensed a quiet complacency engendered by an awareness of the impregnability that the isolation could offer. A quite natural reaction but not necessarily in the interests of all on board.

Management on the beach varied enormously in its approach to its platforms. In some instances the ever-present concern of senior management was palpable, but in others there was a very evident remoteness; with platforms unvisited unless production was dropping or there was a serious incident. In some instances the decision-makers would make a point of visiting platforms if there were to be major changes to company strategy and there would be a face-to-face question-and-answer session. Likewise if there were a tragedy offshore, the scene of the incident would be visited very quickly. But this was not always the case and, sadly, in some instances the senior executives were faceless and unknown. All this was noticed, and noted, by those offshore who had a wide knowledge of different platforms and the conditions on them. More alarmingly there seemed to be general agreement on those deemed to be the most accident prone. It tended to be disconcertingly accurate and in my travels I preferred not to know the latest listing.

Food is not just a necessity, offshore the mess room provides the breaks in a long and intense day. When the price of oil was particularly high it was reflected in the daily offshore catering allowance. Hungry men were confronted with lobster, smoked salmon and filet mignon. It sounded tremendous, the tabloids went to town on the menus and the Good Food Guide must have begun preparing an offshore supplement. But the excitement was short-lived. It was soon obvious that hungry, weary men prefer simple substantial food. The sort that mother makes – or used to. No matter what else is on the menu, the ubiquitous steak is always available, and mince, mutton pies and fried fish are staple fare.

Talking and listening are not the most calorie-burning activities, and I found I had to exercise a lot of self-discipline or no survival suit would have been able to contain me. The wonderful varieties of bread baked on board went un-tasted and, staying with fruit, I felt smugly virtuous and managed to put up with the jokes about spiritual food and nourishment. Breakfasts were quite a feature, with a choice of every cereal known to humankind, and a selection of milk from limp blue to thick yellow. Fresh fruit ignored the seasons, and a full breakfast, known in some circles as 'the works', sat on the hot plates as a constant temptation to which many happily yielded. Lunch and supper/dinner were equally robust and there were times when the galley would go positively cosmic with 'special nights'. The best dishes from Italy, Malaysia or China would be enticingly presented but, it has to be said, that although the mutton pie never won the beauty contest, it often took the popular vote. These special meals broke the monotony, if such it was, of fish, stew and splendid puddings – real puddings and not just desserts. Gastric juices were in full flow and it was little wonder that great attention was paid to healthy eating.

From time to time a nutrition expert would arrive to reveal to the lads the path to gastronomic righteousness. These talks were invariably well attended, but I could never quite work out whether the interest was in personal nutrition or in the lecturers themselves. Invariably the speakers were attractive – the sort who could even make a survival suit becoming – living examples of the gospel they preached.

On the Norwegian platforms the cafeterias remained open for twenty-four hours. In the UK sector eating times were designated, which was just as well, for the variety of food was much better. Even so, it was possible to eat a comprehensive meal at regular intervals round the clock. On a few memorable occasions I would be talking with a group and someone would suggest that we should have 'a wee bit of supper'. I would find myself at midnight rather primly making a selection of cold cuts – my companions were less inhibited. The recollection of my gluttony still makes me embarrassed.

Christmas Day was show time for the galley with the opportunity for the chef to display his talents. It is a semi-working day – more marked by the consumption of vast quantities of traditional fare than of oil production. One Christmas my table companions belonged to a drilling crew from Germany for no matter the special day, drilling, once under way, never stops. Little time was lost and social chitchat was minimal. The conversational contributions were limited to social necessities like *danke* when food was passed and *gut* when food was eaten ,but in the brief pause before the pudding my table-mates, eager to broaden the chat, enquired *Hamburg Wimin Gut?* Their pud arrived before my reply and they were otherwise occupied. I sat back savouring a wonderful first course when I realised that the rest of the table had vacuumed up their Christmas pudding and were en route to the pipes and grease of the drilling floors. Gastronomic endeavour didn't end with lunch. Festive fare prompted prodigious appetites and in the evening the mess room would be full for the cold buffet laid out on a table, framed with birds and fish sculpted in margarine. The selection would have graced a five star hotel.

No doubt describing food in such detail can seem tedious, but it was important. Not just to nourish but to colour and to enliven what could be a predictable, monochrome, existence, and it was a major factor in the maintenance of morale and efficiency. There are times when I wonder about the significance of an entry in my offshore log where, after a visit, I recorded that the worst food I had encountered was on Piper Alpha a few months before it disintegrated.

It was not so much the catering allowance that produced the good result as the morale of the catering staff and their attitude to the raw materials to hand. There were many chefs who were very proud of their galleys and relished the challenge when there was a reason to produce something special.

One day I found myself enjoying an underserved treat when I sat down to a very special lunch. A young couple had made the highest bid in a charity auction and their prize had been a trip offshore. This time the catering staff really went to town. The catering company pulled out all the stops and sent out special cutlery, china and glass-ware. The table was beautifully laid with printed menus and place names. To cap it all the Camp Boss appeared in a black tie and dinner jacket and treated the guests, and the 'hangers on', to a spectacular display of the art of the maître d'hôtel. My vivid memory is not so much of the splendid menu, but of an OIM trying to pretend that this was the way things were always done. Eventually, two happy and extremely well fed visitors departed for the beach with a clear understanding of life offshore that was 100 per cent inaccurate.

The Brent Spar, eventually to be targeted by environmentalists when the time came for her disposal, was an ungainly structure – not unlike a gigantic, slightly fatter Nelson's Column. She stored oil. (It's strange how the Spar is the one installation I have felt I can comfortably refer to as 'she'.) The crew of twenty were accommodated above the oil and directly beneath the small helideck. The living quarters were shaped like slices of cake. In any sort of sea, the motion could best be described as 'interesting', for she was moored to the seabed by six gigantic anchors and chains. As the swell rose and fell and each one took the strain the whole structure would be pulled in a different direction. The motion was rather like a three-dimensional waltzer at a funfair – going up and down, backwards and forwards and sideways at the same time. There were times when I was glad I was wearing my wristbands that prevented sea sickness by exerting a steady pressure on those points that control the balance of the inner ear. It was my misfortune that they were not available when I was in the Navy. There was a small crew and everyone was able to eat together. Each morning the chef would post the day's selection and everyone would tick the items they would like for dinner or supper. There was no choice at breakfast. It was assumed that everyone would eat a hearty meal.

The Spar had the advantage of a small ship with the feel of a united family. Many of the crew had served together for many years, and through the passage of time it seemed to have become an installation where attractive eccentricity could flourish. The wireless operator – the quintessential radio 'ham' – would spend happy hours devising an even more complex system of aerials for his Yorkshire garden in his quest for perfect wireless reception. A deckhand was a keen student of the writings of the great philosophers. He did much for my humility, not least when he wanted to discuss the 'I–thou' relationship developed by Martin Buber. After his verbal drubbing any intellectual self-confidence that was left would be annihilated by the chef who was the ardent compiler of general knowledge competitions and a quizmaster of fearsome authority. Once, when I innocently queried his answer to the question, 'Where was Princess Margaret born?', I nearly caused the nearest thing the North Sea was ever likely to get to a riot.

When the first platforms were designed in the 1960s there were no precedents on which to model the layout. The prime concern was the successful extraction of oil and the structure had to be such that this could be an effective round-the-clock operation. The fact that people were involved, whilst not exactly an afterthought, does not seem to have been much of a priority. The early models – the Brent Spar provides an interesting example – were the equivalent of the Model T Ford. The tendency was to build vertically. This meant that steel ladders and stairs were a predominant feature. As experience increased and more was learned about offshore requirements, bits and pieces would be bolted on. They were unattractive constructions with little discernible logic to their layout. It could be disconcertingly easy to get lost, and as no-one ever wanted to hang over the rail admiring the view, preferring to remain behind their closed, look-alike heavy sliding doors, a platform could become like the *Marie Celeste* with no-one around to give helpful directions.

When I got lost I always felt remarkably foolish, and it was quite a relief to be able, eventually, to slip into the PLO and back into the silence and welcoming warmth. Moving around the platform was made even more difficult by the constant alterations. These could mean extra sleeping accommodation, increased storage space (always at a premium), or additional electronic equipment, and it all could add to the possibility of disorientation. Being lost on a platform – how absurd it sounds – did you no credit. If an emergency were to occur, in an instant you became both a liability and a distraction. On the few times it happened I would console myself with the lame explanation that I visited so many different platforms that I could not be expected to remember every layout. Not a good enough excuse, but it did seem to me that, with the mobility of the contract work-force and the unending quest for greater safety, there was an argument for a greater rationalisation in design.

Being so peripatetic myself gave me the opportunity to make comparisons between platforms, and it opened my eyes to ways in which living and working in the North Sea could be improved. From the engineering point of view the passage of time has shown the platforms to be of proven robust construction, and a huge concentration of wisdom and experience made sure that the oil was extracted with efficiency. However, there were fewer insights into the human requirements. My profoundly non -technical eye would look over the living and eating accommodation and soon pick out areas that could be improved. Cabins were marked by their lack of storage and shelf space – made more obvious once women worked offshore – and the communal toilet facilities brought back vivid memories of life on the lower deck. As the years passed, the development in offshore platform layout became evident and more thoughtful designs began to appear. To my untutored eye not only the living quarters, but whole platforms, seemed to be more rationally laid out with a larger, uncluttered deck area. Less of the vertical and more of the horizontal in construction simplified maintenance.

By the late 1980s the profile of the structures had fundamentally changed. They looked less clumsy and were much easier to get around. A major plus factor was the natural daylight that flooded into the living quarters.

Unsurprisingly, the engineering complexities continued to be a mystery to me. It was the norm on my first visit to a platform – it must have happened at least fifty times – to be taken on a guided tour by a nominated enthusiast. I think it was assumed I would be able to get a grasp of the platform layout, but the tour invariably gave my guide the opportunity to wax eloquent about 'christmas trees', the gigantic control valves that prevent a blow-out, and 'pigs', immortalised by James Bond when he escaped from the other side of the Iron Curtain in a steel capsule expelled down a pipeline. But that was an 007 excess and they were used, not as a human postal service, but rather more prosaically to keep the pipelines clear. At first, it was all very interesting, although incomprehensible, but after the umpteenth tour, I realised that I lacked the basic enthusiasm of the committed engineer. If anyone ever again mentions 'cascades', my eyes will glaze over and if someone begins to rhapsodise over psi's (pressure per square inch) the legs begin to weaken.

There was a good working relationship with the Norwegian offshore chaplains (all five of them!) and I was able to visit quite a few platforms in their sector. The Norwegian design concept was quite different, with a huge accommodation module set on its own platform removed from the production modules. Up to a thousand crew would be accommodated and would use linking bridges to walk to their places of work. Norwegian legislation forbad accommodation being provided for more than two persons per cabin. In the UK, with its less generous design concept, there had been a history of four-man cabins, with the resulting pressure on space and privacy. Happily conditions improved and single/double cabins with en suite shower and toilet facilities became the norm. But it was very plain to me that it was personal attitudes, and not the most contemporary living conditions, that made a 'good' platform.

Some crews were outstanding in their determination to make the best of their lot and this was typified by the most venerable, yet still active, platform in the North Sea. Whilst quietly pumping away, it always stood out with its particularly happy atmosphere. The Auk was small, and did its best to look attractive. The expression on visitors' faces was memorable when they climbed down from the helideck to be greeted at a turn in the steel ladder by a splendid display of tulips cultivated under glass. Accommodation was tight but it was an outstanding example that 'small is beautiful'.

The Hutton TLP (Tension Leg Platform) was remarkable for two things. Alone in the North Sea it had special rules for billiard players – as the legs adjusted to the swell on occasion, the movement meant a shift in the snooker balls. This could produce critical situations. Hutton also produced the only 'in house' magazine I came across. It was called *The Oily Rag* and seemed to be modelled on *Private Eye*. It had an editorial board blessed with vivid imaginations and it was compulsive reading in a North Sea never wanting for gossip with news travelling around platforms at lightning speed and ever-increasing inaccuracy. Doubtless UK libel laws did not extend outside territorial waters!

The opportunities for less sedentary activities were very limited and required ingenuity. Golfers' driving nets were not unknown, but their location produced a hazard not anticipated by the R & A. They were usually installed directly below the helideck, fine in normal circumstances, but the arrival of a chopper could be devastating to the back swing.

At one time on the Fulmar there had been a badminton club. Games were played in the hanger that had been constructed for a helicopter which, in the times of plenty, had been permanently based offshore. The hanger came into its own when converted into a television studio for a Christmas Eve service that was transmitted live to the beach – the only live TV broadcast ever made from offshore.

Most platforms had room for a gym, often rather small and a bit 'sweaty' and in every crew there were the disciplined who went through their daily work-outs. Overall there were many who were wise enough to realise that renewal for another day needed more than sleep.

Entry into the PLQ was marked by the hiss of the air pressure stabilising once the double doors were closed. Crossing the threshold was to enter a peace and quiet that could be overwhelming after the unending thump of compressors, the screech of metal on metal, and above everything else, the relentless buffeting wind. For this was the everyday environment of most offshore workers as they went about their business; anonymous figures in hard hats, earmuffs and protective goggles and wearing those insulated overalls that do their best to create a workforce of Michelin men. Outside you had to shout close to someone's ear to make yourself heard. Inside the PLQ was a transforming experience; warm and with no need to raise your voice. Short-sleeved, open-necked shirts were the rig of the day. The contrast was striking, but there was a downside to this haven. The controlled atmosphere could result in dehydration and dandruff, dependent on air conditioning, and there could be an absence of natural light in the cabins. When one could glance out to sea it could be unnerving, for one gazed at restless waves with the sound completely masked by triple glazing. Often the unnatural stillness reminded me of a recording studio. This absence of noise could be a bit strange. It was rather like living in a capsule and certainly far removed from the clock watching, time-controlled environment that we accept without question in our own shore-bound lives. There was none of the 'Thank God it's Friday' syndrome, for the rhythm of work was relentless – each day resembled the one before and would be the same as the one to come. It was a predictability that could be disconcerting when talking to someone who knew exactly which Christmases and anniversaries he would miss in the next few years.

On the older platforms cinemas had been incorporated in the PLQs but with the arrival of cabin TVs it seemed from my own experience they were rarely used. I could not help thinking of the provision that would be made for service personnel in not dissimilar conditions, and the stimulation that resulted from visits from celebrities if they could be encouraged to go offshore. They could do so much quite simply, and it was not a performance that was required so much as a talk with a question-and-answer session that promoted discussion. The keen and lively minds offshore would relish contact and shared insights from acknowledged experts in just about any field. It's strange that the imaginative deployment of gifted individuals to encourage and support their fellowmen has to be prompted by a war.

The 'quiet room' usually housed the tranquillising fish tank. The room was not overused but, like many important things, its value was in 'being there' and more of the crew were grateful for its existence than is probably realised. I found it a useful place for 'the putting the world right' sessions which were always marked by a frank exchange of views – and a heartwarming inconclusiveness when it was time to break up.

Whenever a trip was over and it was time to leave a world buffeted by the wind and the noise of machinery, I always felt I was saying farewell to a unique way of living. For some reason, it made me think of Star Trek, but there was one big difference. This piece of machinery was going nowhere.

6

The Commuting

LOG EXCERPTS

Uneventful fixed-wing flight to Sumburgh. Sat beside the personnel expert on pensions, an attractive girl, on her way to North Cormorant. It turns out that being 4 feet 10 inches she had a special mini survival suit waiting for her.

Cormorant Alpha is an unusual platform with its own air traffic control centre manned by controllers who are employees of the Civil Aviation Authority. It was an eye-opener to get some glimpse of the complexities of helicopter traffic in this remote yet crowded area.

I question whether the psychological approach is of the best when going offshore. The procedure is (a) fill in a next of kin form (b) get into a survival suit and listen to a survival lecture marked more by its optimism that its reality and (c) get in the chopper with earmuffs that cut off your contact with the outside world and leave you, dangerously, alone with your thoughts.

A good flight in the helicopter to Sumburgh. Driving rain but none of those headwinds that can make the flights seem interminable.

I distinguished myself at Brent Alpha where I disembarked. This caused a certain confusion with the chopper crew because they had no-one to take on to Brent Charlie. Apparently most people, at some time, make this mistake.

It's very easy to do! Indeed, I was to have an experience when all the passengers disembarked on the wrong platform.

⚲ *The chopper had to be replaced and we sat in our survival suits in a crammed departure lounge for over an hour. It was a hot day and was something to be discouraged.*

⚲ *Things are changing. Computerised check-in means that all the form filling previously required i.e. next of kin, blood group etc., is no longer needed. Not only are things much simplified but psychologically it helps towards a calmer frame of mind.*

⚲ *Two crane accidents in one week. In one the main beam snapped, providentially in the one area where it could do little damage (it had just completed a lift of 9 tons of aviation fuel). In the other incident the wire parted, but fortunately it jammed and the load was not lost. There is so much luck about accidents.*

⚲ *Took off at 0845 hours. The Shetland Isles looked glorious in the bright sunshine – great acres of golden sand and crystal clear, deep blue, water. After ten minutes the flight was aborted due to a technical failure, 'Don't worry chaps, it is only a minor electrical fault – honestly!!'*

There are many descriptions of an Aberdonian but one of the shrewdest is of an individual who never looks up when a helicopter flies overhead. The almost incessant sound of rotor blades strikes every new arrival at the main airport, and the distinctive 'chug chug' has identified well defined routes over the city for decades. Inevitably, the passage of time has anaesthetised locals to the sound and they never spare a glance for the aircraft, let alone a thought for the survival-suited passengers who are either going to, or returning from, their strange workplaces over the horizon.

Whilst the chopper is so much a part of the sight and sound in the oil capital, the number of people who travel in them is small, and confined to those directly involved in work offshore. The chopper, the North Sea oil platform and the UK economy, make an intriguing trio around which has developed a unique mythology. As a visitor there is already something mysterious that surrounds those lonely steel islands before you start your journey and this is further heightened by the way, the only way, you get to them. When photographs for public consumption are released about the offshore oil industry they are entirely predictable, and tend to appear in the business section of the newspapers devoted to the UK's oil prospects. Only rarely is there pictorial evidence that people are involved. Occasionally, when there is an article with some human interest, the reading public usually have to be satisfied with a photo of a crocodile of individuals filing out to a helicopter enveloped in their sexless, anonymous, survival suits. In fairness this impersonality can be explained. A scattered workforce, converging, from all over the UK, and beyond, to security conscious heliports, and then being whisked away to platforms does little to encourage the emergence of individuality, and if a casual observer were to fly over a platform the apparent absence of human activity could be astonishing. Only when an incident takes place is there a seismic shift in attention. Then the media replaces technology with humanity and the workforce becomes the centre of attention.

The offshore mystique is made up of a heap of ingredients that can stir the imagination and ignite the curiosity. Put them together and the result is an intriguing picture. Take a survival suit; a helicopter; a mysterious, isolated, destination; add a dash of bad weather; flights over forbidding seas and then, for good measure, throw in a slug of darkness, for the northern latitudes offer few hours of daylight in the winter months. In this cocktail there are hints of hazard and suggestions of unpredictability, and this can fuel the imagination of the TV viewer or the tabloid reader.

For much of the year the northern North Sea is inhospitable and empty, and the all-enveloping survival suit has been shown to be much more than a practical gesture to safety. On the rare occasions when there has been a ditching, and escape has been possible (thanks to well rehearsed survival training), the suit has countered the perils of hypothermia. However, there is more to it than the obvious. It has several incidental uses, not least in acting as a psychologically reassuring cocoon into which one can relax when seated in a chilly chopper on an inhospitable night. Again, it's a first class protection against the elements when crossing a storm-swept helideck. But there can be a downside. On a hot summer's day when the sun streams through the cabin windows and zips are slipped down to a vaguely indecent level, they can provide a good imitation of a sauna.

From the moment they first wriggle into their survival suits, most newcomers to the offshore scene are on a 'high'. The novelty of the whole experience can be overwhelming. Then after a few trips, quite suddenly and deflatingly, there comes a time when the procedures become routine. The queueing for the suit in the heliport departure lounge; the muttering to the keeper of the wardrobe the size that fits you and that you have come to know so well; the removal of the shoes and the energy-sapping wriggling as you make the suit comfortable. Being particularly anatomically imperfect

– I prefer the description 'stocky' – I soon discovered that my breadth of chest was usually associated with someone a few inches longer in the leg. This meant that whenever I put on a survival suit the legs looked as though they had just come out of an accordion. But there were compensations in the looseness – I could always store a newspaper and a paperback in the thigh pocket.

Once everyone was suited up they had to put on their life-jacket. At one time this had been a puny object tucked into a pouch and secured round the waist by tapes. Then came a much more robust and sophisticated inflatable waistcoat, secured by straps as though it was a parachute harness. Once passengers were fully dressed, there then came the metronomic delivery, by a member of the despatch staff, of the necessary action to be taken in an emergency; pulling up the hood; putting on the gloves, ensuring all zips were fully closed. The latter usually meant the near strangulation of the passenger 'chosen' to act out the instructions. The forehead light would be clamped on (thanks be to God for the inventor of Velcro) and the whistle would be located. All of this was in such contrast to a commercial fixed-wing flight with its passengers manifestly uninterested in the safety instructions.

Going offshore was for real and everyone recognised it. No matter how frequent the travel, close attention was paid, and was expected to be paid, by all present. There was no patience with indifference for there had been times when the training had been put into operation in real-life situations and it had been shown to work. This said much for the Survival School in Aberdeen which had been such a pioneer in helicopter passenger safety.

The boarding routine never varied. A single file (in bad weather with hoods up, making it look as though a host of monks were being transported by 'Conair') would shuffle out to the chopper. Each passenger carried the single permitted holdall, soft and squashable, and easy to stow in the luggage hold. Like churchgoers, the regulars, once on board, always seemed to have their favourite seat. I knew this and always opted to bring up the rear of the file, but there was a downside to this self-sacrifice. Usually by the time I boarded the chopper, the only vacant seats were in the rear of the aircraft – where the pull out windows were the smallest. Comparing my slender form to the size of the aperture I could not hold out much hope of a speedy exit.

When I first went offshore, passengers put on earmuffs to deaden the noise. An excellent provision and the peace was blissful but it did mean that everyone on board became insulated from all communication. With the introduction of plug-in headphones it meant that one could hear what the pilot had to say . . . well, most of the time, but some pilots lacked a clear microphone technique. It was expected that at all times the volume was turned up sufficiently to allow one to hear any instruction. In between announcements and information, there was music to while away the hours. Passengers were encouraged to bring their favourite tapes and hand them in to the cockpit on boarding. The ever-versatile second pilot would take on the duties of DJ, announcing a selection of music that could be infinite in its variety – you turned up the volume for some and down (very) for others.

The Sikorsky 61 was the old war-horse of the North Sea. It was slower than its sleeker more modern rivals and much nosier but I preferred it. I felt that travelling in the more modern Bells and Pumas was rather like being transported in a cigar tube. When full of men and women suited up, it could be very claustrophobic, and I had a sneaky feeling that all the instructions about evacuation were a bit theoretical.

No matter what technical aids were available, at the end of the day, it was the pilots and their skill upon whom everything depended. They were a remarkable group with a huge flying experience gained all over the world. Some had been 'crop dusters', others had been in the Army Air Corps and the RAF and there were veterans of Vietnam, Forestry in Canada and Air Taxis wherever they were needed. The strain that went with the job was revealed in their faces; many looked far too old for their years – curiously, the real veterans seemed to acquire the sort of complexion that can be found amongst those who do not have much exposure to fresh air. Amongst ballroom dancing aficionados it used to be called a 'Mecca tan' after the dance halls of that name. Air crew were always thoroughly professional and courteous in their dealings with everyone, yet strangely remote. However, when they landed a new crew on a platform and wished them 'a good trip' you felt they meant it. They spent little time on the platforms and when they did it was not easy to get to know them. Maybe they preferred it that way.

One day as we took off at Sumburgh, I looked along the length of the cabin – never difficult from my usual seat at the rear of the chopper. I noticed a particularly slender chamois gloved hand holding the collective control. Some hours later when we disembarked a very feminine voice wished us farewell. There were a few women who captained helicopters but apparently some passengers were unhappy about women pilots and the problem was ingeniously solved. You rarely saw the face of the pilot. When boarding you entered the side of the aircraft and made your way up to the passenger area, so only the voice would be the give-away. The second pilot (male) would deliver all the preliminary announcements and any en route instructions. It was left to the female's dulcet tones to deliver the fond farewell and confound the male chauvinists. I had already discovered to my surprise that I was afflicted with a mild attack of the same condition. One day after our fixed-wing flight arrived in the Shetland Isles, the female captain stood at the exit saying goodbye. I made some stupid remark about the novelty of being piloted by a woman and a very large man beside me immediately and unanswerably responded, 'And what's wrong with that?' Total deflation of the chaplain who was supposed to love all his fellow men and women.

Offshore operations are conducted at a very northerly latitude, with many flights taking place in darkness. All too often the weather could be stormy with the chopper having to push hard if there were a headwind, but there was the compensation of nights of startling beauty with the Northern Hemisphere cloudless and the constellations twinkling away in their glory. And there were unforgettable moments in the summer when the sun blazed down on a limpid sea and there was only a hint of creamy foam as the swell rose and fell against the platform's legs, and the stark architecture – all straight lines and right angles – was softened by sun and shadow.

Sea mists were another matter. At one moment it could be a gloriously sunny day and the next everything would be shrouded in a dense, chilly, blanket that reduced visibility to zero and magnified all sound. At times it seemed as though the platform was in the middle of an echo chamber. Everything became cold and clammy, and mildly claustrophobic, and it needed a conscious effort not to whisper. Interestingly, the variations in weather could be very sudden, and very local, and could result in major logistical problems.

It was possible to set out from Aberdeen in brilliant sunshine and having reached Shetland be trapped in an impenetrable haar, that wet cold that can be a feature of a North Sea summer. Weather and helicopter flights in the North Sea have an uneasy coexistence. When air movements have to be suspended there can be a domino effect on crew changes. Those who have completed their couple of weeks and are expecting to go home have to endure the frustration of returning to their usual routine, knowing it has to be maintained until their relief's arrival – whenever that might be. It was equally irritating to the relief crew whose offshore trip was now in limbo. However educative an enforced stop at Sumburgh might be, they just wanted their trip to begin.

The one hotel near the airport in Shetland was known to all as 'Fawlty Towers'. In fairness to the establishment it must have been very difficult to run. For most of the time it would be accommodating holiday-makers or enthusiasts working on the nearby archaeological dig. Come the mist came the deluge. Suddenly, there would be an invasion of dozens of oil men trapped indefinitely, but expecting to be housed, and fed, yet liable to depart as suddenly as they arrived if there were a break in the weather.

One day the weather had been peerless in Aberdeen, but on arrival in Shetland the visibility suddenly dropped to zero – the chill factor was almost as low. We had to stay overnight at the celebrated hotel but we were in luck because our enforced stay coincided with a 'shorts and shades' evening in the bar. The local populace found their way through the gloom dressed in styles that would have put the loungers on Malibu Beach to shame. Needless to say the oil industry, despite a lack of vacation gear, entered into the spirit of the evening. It may have been bleak outside, but indoors wearers of Hawaiian shirts competed for the 'loudest' prize. The beer flowed sensibly, but just sufficiently for inhibitions to be gently released. 'Andy, I'm pissed but you'll not take offence,' was the preamble that would lead to stories either comic or sad, sometimes both. It helped me to understand yet more about a strange life and its stresses and strains.

You need a large-scale map of Shetland to locate Scatsta and Unst. To the oil industry they were an inseparable part of their work. For those employed at the vast oil terminal at Sullom Voe the journey to Scatsta was but the beginning of a blissful journey home. It was more a landing strip than an amenity-rich airport, but it served the terminal and was a vital link in the communication network that threaded through the offshore operation.

At one time Unst, the most northerly Shetland Isle, boasted a post office. But its claim to fame could be further boosted as the site of the ultimate in isolated airports in the UK. Some of the oil companies made use of it because it shortened the helicopter trip to the Shetland Basin and they would fly their crews from the South in fixed-wing craft and then transfer them on this lonely runway. Its air traffic controllers were an interesting bunch. In normal circumstances, they were far from overworked, yet always indispensable, and in an emergency they were absolutely pivotal. Their particular trips lasted for several weeks at a time before they returned to their homes for some leave. 'Home' for them all was the North of England and I often thought of these diverse communities and the contrast with a work location where, for days on end, they would be looking out on a wet, windswept and often deserted runway. But they all liked it and something of the magnetism of this very singular outpost of the British Isles appealed to them to such an extent that some even decided to up sticks and live permanently on the Islands. They were not alone. Over the years I came across a surprising number of oil families who had settled in a land where tranquillity, good neighbours and a positive sense of community exerted a pull that had proved irresistible.

The realisation that the disadvantage of a rigorous climate can be comprehensively outweighed by the warmth of supportive neighbours, has also been understood on the Orkney Isles, not least by those who worked at the Flotta Oil Terminal who had a complicated journey to work involving an Aberdeen to Kirkwall flight, a bus to Orphir and a ferry across Scapa Flow to the island whose gas flare was the only indication of another enormous installation.

Both Flotta and Sullom Voe were fine examples of huge engineering projects which could be skilfully crafted into the landscape and then operated with such sensitive attention to conservation requirements that only a nitpicking environmentalist could object to their presence. The planning, and its realisation, was so successful that I found a certain irony in the fact that an already invisible offshore industry managed to make its landward operation equally anonymous. It suggests a very effective formula. Put together a remote place and a sensitive design and the spotlight is swung away from a crucial industry.

The offshore family is as invisible when it gets to the beach as it is in the North Sea. Commuting to work has starting-points all over the UK and beyond. Some workers travelled from France and Spain and they reckoned their journey could take less time than that faced by their colleagues in the South of England. One man married to a Frenchwoman remarked how he caught the TGV at Cluny in Burgundy to Paris and then caught a plane to Aberdeen – his journey could be remarkably speedy.

It never ceased to amaze me that a resident of the Isle of Wight could be so much more familiar with the Shetland Isles than an indigenous Scot. The Isles' beauty and isolation had a deep appeal for some. Although the offshore worker was a bird of passage, and for many Shetland meant little more than the airport at Sumburgh, there were those who came from more southern climes, not least those working at the oil terminals, who fell under the spell of the islands and became enchanted by the way of life with its repose and timelessness.

Those who set up their homes in the far north played a full part in their local communities. Talking with them was always informative, for they had found a way of life that was genuine. Fascinated by the elemental powers of the storms, and struck by the emptiness of the landscape, often to their surprise they felt thoroughly at home. Some admitted discovering within themselves resources of independence and initiative that they had never suspected. However, not everyone felt that way, and there were many who just could not wait to return south to what they understood to be civilisation.

The longer I worked offshore the more my own commuting became a bit of a nightmare. Thanks to the goodwill of the majority of the oil companies I was permitted access to most of the platforms and my wallet bulged with a heap of plastic passes that allowed me offshore. Anyone giving it a cursory glance probably thought I had a mania for collecting credit cards. My personal difficulty was in the resolute independence (stubbornness would not be an entirely inappropriate word) of some companies who persisted in having their own distinctive routine for going offshore. The cost must have been considerable and seemed so unnecessary, but once the economics of the operation were more clearly examined in the light of fiscal reality, much of the resistance to co-operation disappeared. Systems are now in place that could not have been contemplated a few years ago. Everything is simplified and more accurate. Most importantly, in times of incident, vital information is now more readily available.

A major factor in all the change that was going on was the unending pursuit of safety in operation procedures and techniques. The heliport safety videos became so familiar that I came to know the commentaries by heart – but I found I never became anaesthetised by the message. However, I did become irritated by the wide range of departure points. At one time there were micro heliports within the heliports where some companies persisted in running their own operations. Such was the individualism that custom-built minibuses would take the new crews out to the waiting chopper, passing en route other crews that would be filing out from the main departure point.

Most companies forsook their precious independence and shared heliports where there was a cafe, a news-stand with daily papers and a wide range of reading material, from the cellophane-wrapped top-shelf magazines to the latest best-selling paperbacks. I imagine the offshore workforce must be one of the best-read industrial armies that has ever existed. The heliport was exactly the same as a commercial airport with all the same familiar procedures; flights being called; security checks enforced. But there were some differences; the impounding of cigarette lighters must have resulted in a definitive collection to which I made a not ungenerous contribution and there were thorough searches for drink and drugs. The suiting up and the rigorous safety briefings were unique.

A marked feature at Aberdeen heliport was that by early afternoon most of the air traffic had ceased and the terminals, in contrast to the dawn bustle, were almost deserted. I have never been one of those people who delight in early rising and there were times when I arrived on some remote platform in the far north and, looking at my watch, thought of all those sensible people still tucked up in bed, and then pondered on the miles I had travelled and the full day that lay ahead. Hopefully, I would be having my lunch whilst people on the beach would be getting around to savouring their mid morning coffee. There was a timelessness in the offshore life that took a lot of getting used to.

As I became more experienced I developed my own, extra, embarkation routine. I would treat myself to the latest paperback and then buy a bundle of newspapers for the crew who had been relieved and would enjoy a good read on their way back to the beach. Significantly, the broadsheets were the most popular. Arriving on a platform was quite a ritual, with those going ashore patiently lining the steel stairway leading to the helideck ready to proceed at the double to the chopper. There was always the need for a quick turn round, not least if bad weather was closing in. It could be a frantic mixture of foul weather, a noisy chopper whose engine was rarely shut down, and the shouted warning about avoiding the rear rotor. Sadly there have been some tragic accidents. Incredibly, it is all too easy to get in its way.

The crew change was important to me and sometimes the descent from helideck to PLQ could be quite slow as I was given snippets of useful information about people. But when disembarking you had to be careful. In the early days it could be difficult to know where one was. From the chopper most platforms can look alike, and if it was dropping people off at different destinations the pilot was duty bound to tell you where you had landed. The name of a platform was painted on the helideck, which may have been useful to the man in the cockpit but virtually invisible to the peering passengers. The situation was not helped if the pilot was disinclined to be chatty and would assume mind-reading powers amongst his passengers that were a trifle optimistic. On one occasion I disembarked, with great confidence, on the wrong platform, and great was the consternation as my baggage disappeared over the horizon. There was an even more spectacular bloomer when I was part of a complete passenger list which presented itself at platform reception who were somewhat taken aback. Instead of eighteen travel-weary humans, they were expecting a spare part.

My own journeys never seemed to be straightforward. Being the odd man out, I would be given a spare seat in a chopper that was taking out a crew change – not necessarily to my destination. Once disembarked, my hope would be that there was a seat on an inter-platform shuttle. These small Bell choppers always had two pilots who seemed delighted to show off their skills, or so it seemed to me, peering through the fingers that covered my eyes, as we dropped alongside some vast and potentially all-consuming flare.

These shuttle choppers were small, and seemed to become even smaller when tightly packed with men returning to their flotel or base platform. Although they flew in most weathers, there were times when shuttles could not return to base and the men had to bed down in the corridors of the PLQ or in the cinema (it had its uses) after being issued with a sleeping bag, and an overnight pack – the sort you can get on planes and trains. For men working twelve-hour shifts these sort of disturbances could be especially fatiguing, yet there were many who preferred being marooned to catching a packed shuttle in foul weather.

The subtle stresses to which crews were subjected were rarely mentioned. Maybe they were all too difficult to articulate. I found that the face, and not least the eyes, would tell its own story of bone-deep weariness. As in every intense activity, the gap between those at the sharp end and those back at base was reflected in the offshore industry. There were times when I was astonished by the lack of sensitivity shown to those who were commuting offshore. It showed in small but significant ways. Delays caused by bad weather were a typical example. If there was a hold-up at the point of departure, there could be a long period when everyone sat around awaiting and there could be total lack of communication. When eventually word did come, it would provide the sketchiest of information, delivered seemingly reluctantly, by people whose body language suggested we were fortunate to be told anything at all. It was possible to spend a whole day at Sumburgh or Aberdeen without knowing what was happening. The first news could well be that everyone would be spending the night in an airport hotel. On one overnight stay at Sumburgh, I found that the stewardesses involved were handed bedroom keys for the annexe where there were only two communal showers and communal washing. It summed up an attitude that was both unfortunate, undeserved and yet not unexpected. The offshore army is like any other army, where differing responsibilities and experience can find it difficult to coexist. The misfits and malcontents, whose behaviour could attract a disproportionate amount of attention, were partly to blame in generating attitudes that excluded the overwhelming majority from the dignity they deserved.

In 1989, the 25th anniversary of offshore operation was reached. Sadly, it is not possible to say 'celebrated', for the landmark went off like a damp squib with scarcely a murmur of celebration or appreciation. A unique opportunity was lost to recognise those who laboured in an industry that is universally accepted as being crucial to the UK economy. I could not help thinking that if other armies can have a celebratory reunion in the Albert Hall, why not the offshore army?

As I travelled backwards and forwards, and lived with these men and women, I found it increasingly difficult to equate my reason for being with them, 'You matter as individuals and you are important,' with company attitudes which so often seemed to express the opposite. One day I was talking to a perceptive OIM and he remarked, 'We are very good at telling people who make mistakes that they are not much use. We have to get our act together and in all sorts of ways reassure and compliment people when they are doing a good job.'

7

Safety

LOG EXCERPTS

Of course, companies and their satellites are not charitable institutions. Their task is to extract oil and make sufficient profit in the process to enable them to plan with realism in a capital intensive activity. They must invest more in people. North Sea offshore oil extraction requires men of a certain basic calibre – no matter how humble their occupation. This 'calibre ingredient' must not be undervalued. If it is, then the production and extraction problems will be magnified; most certainly the safety factor will become of great concern.

The trip to an ageing platform that was not subject to production pressures has made me reflect on the atmosphere that will exist in the North Sea, presumably in the next century. It will be a time of testing, notably for safety.

Went to my lifeboat station and had instructions in launching and starting up the boat's engine. My instructor, Mervyn Jenkins, was a delightful, but verbose, Welshman (living in Southern Ireland). I couldn't take it all in so am fascinated in what I might do as the sole survivor in an emergency. My hunch is I would coax the thing to fly – or else I'd walk on the water!

It was the summer of 1988 and one of those balmy evenings when the wind is a mere zephyr and the sea like a moving mirror. The support vessel *Tharos* was ugly, but practical, dominated by her huge crane as she lay alongside what little remained of the Piper Alpha Platform. The noise from all the fractured pipes was deafening, a compelling reminder of the elemental forces that were now escaping uncontrolled. The release of gas, under immense pressure, had ignited the fractured pipes which stuck out of the water like a row of rotten teeth. I had flown out in a helicopter crammed with floral tributes to lead a memorial service for all who had perished a few days before. All of the salvage and control crew who could be spared had gathered in the well deck of the vessel. Many had lost friends; the prayers helped to release some of the emotion, as we remembered before God those who had died, those who had survived and their loved ones who mourned on the beach. The service ended and the silence from the group was total as the floral tributes were gently lowered into the sea. An immense multi-coloured carpet slowly drifted out of sight on the strong current. Then in the stillness came the haunting sound of a lament, played by a member of the crew. It was one of those times when there was a 'tingle' factor as the pipes so uniquely expressed the feelings of men in their distress.

On the *Tharos* there were a precious few moments when everyone was able to be alone with their own thoughts and with the time to regain their self-possession before returning to their hard tasks of control and recovery. In the centre of the group I noticed a small, portly, man wearing a moth-eaten pullover. He looked somewhat elderly to be offshore, not least in these horrific circumstances, but he was obviously at ease in the situation. This was my first sight of the legendary fighter of oil fires, Red Adair, who had crossed the Atlantic to control, and then extinguish, the flames that had been roaring away for several days keeping the remains of the platform white hot. Whilst the man I met was rather less physically impressive than John Wayne's film portrayal, he made up for it in his forthright conversation. Having expressed surprise at meeting a 'Reverend' in such unusual circumstances, together we watched in sadness the patchwork of flowers disappear over the horizon. Suddenly he burst out, 'Who, but a nutcase, would place a factory, a hotel and a heliport on top of a wellhead.' Insane or not, that is precisely what had happened since the first North Sea platforms were installed.

From the outset there was an understanding, and acceptance, of the dangers that abounded; just as there was recognition that only ceaseless vigilance could master the inherent risks. It was recognised that intensive training and comprehensive legislation would be needed to do everything possible to ameliorate the hazards. The North Sea venture was a novelty: everything was new and the absence of precedents meant that the demands on men and machines were unknown. The only way to learn was to get on with the job, with sufficient inbuilt flexibility to respond to situations as they occurred.

Take helicopter travel – it is an acquired taste that has no appeal to some people. A chopper's manoeuvrability makes it unique, but most of those who fly in them are acutely aware of the lack of wings and the crisis produced by engine failure. Choppers have a modest gliding capability and fail spectacularly when trying to imitate our feathered friends. To make the situation yet more deterring, the North Sea is cold and uninviting – hardy bathers on a summer's day will testify to that. It is noticeable that East coast holiday resorts, in the days when they were fashionable, were usually described as 'bracing'. But true to the traditions of oil exploration, it was this bit of unappealing ocean that concealed the black gold that had to be developed. Appropriate training was devised.

Everyone who travelled offshore was required to go through a basic survival course that included the simulation of a helicopter ditching at sea. A mock fuselage would be lowered into a specially constructed pool where wind, waves and darkness, all with appropriate sound effects, could be created. The fuselage would be capsized under water, and the survival-suited passengers had to unstrap themselves, escape via the window frames, surface, inflate their life-jackets and clamber into a rubber dingy. The courage and determination of some of those seeking a survival certificate was remarkable. There were some who, although they could not swim, were prepared to subject themselves to the ordeal and passed with flying colours.

Incredibly, in the very earliest days, people went offshore in suits and city shoes, but mercifully things soon moved on. Life-jackets became more robust with the flimsy garment still beloved by airlines being replaced by the waistcoat style that was secured over the survival suit. For the novice the journey offshore could be quite fraught, and the sense of adventure was not reduced when, having endured the novelty of the journey to the platform, the passenger would land all too aware that his destination could be highly inflammable. However, his confidence would not necessarily be eroded because he would have been through a course in basic fire- fighting and escape from smoke-filled areas. The eventual introduction of smoke masks in the cabins was another confidence booster. Quite apart from the practical value in all of this, but equally importantly, there was the boosting of confidence. Above all else, the procedures rammed home the fact that safety was the top priority.

Every platform has a safety officer responsible for the observance of regulations and the implementation of secure routines. Not least, he has the dangerously repetitive task of taking all new arrivals through the platform's well tried safety procedures.

It always struck me that the key to effective communication, which could mean the difference between life and death, lay in the freshness of approach. When there was ingenuity and sparkle the message got home, but at times the induction could be stiflingly dull and guaranteed to produce inattention. This was not helped on one occasion when the video player screened a film that frequent projection had made almost invisible. It was hard to believe that the exercise was intended to be taken seriously.

On one memorable occasion the safety representative delivered a friendly and enthusiastic briefing. However, this positive side was rather cancelled out, and yet made peculiarly compelling, as he manfully tried to cope with a major speech defect.

In my own particular case, the biggest problems were the variations in safety procedures from one platform to another. Alarm systems could vary, with the sequence of sound and warning lights producing infinite combinations, and there appeared to be no common mind on the location of muster points. On some platforms lifebelts would be found hanging on the back of cabin doors but, just to complicate things, on others they would be located in vast steel chests beside the lifeboats. Problem: to find your lifebelt, first find your lifeboat. To my already confused mind the operation of the boats themselves was the ultimate mystery.

One day a safety rep, who seemed to have little else to do, or else he had a somewhat cynical sense of humour (maybe a bit of both), led me to a lifeboat. We climbed on board and very soon my head was reeling with all the instructions I was supposed to remember, and even worse, to implement if I was to successfully start up the lowering mechanism and launch the boat into the sea below and then sail it safely away from, presumably, a rapidly disintegrating platform.

This 'hope not' scenario left me in a bit of a daze and I found my thoughts straying back to my first months in the Navy and a grizzled old chief petty officer who sat me, and the rest of my entry, in a whaler, fortunately on dry land. Suddenly, he pulled out a massive bung in the keel that made it possible to drain the boat after a sea trip, and waved it at a group of mesmerised youths saying, 'Now we mustn't do that when we are afloat must we?'

I tried to take everything in, but with mounting incredulity. The instructions I was listening to seemed to be based on two highly optimistic assumptions. Firstly, that the lifeboat had been successfully lowered some eighty feet into a heavy swell. Secondly, and even more grotesque, was the idea that having reached sea level I would be sufficiently self-possessed to start up the engine and sail away a safe distance. Then, apparently happy in a survival suit (if I had found it), I would calmly wait to be picked up. For me that presented another problem because he was not able to tell me how to get out of the wretched thing once all its hatches were sealed and it was totally battened down. Anyway, I do not think it mattered because by that time, and in the likely motion in a hermetically sealed boat, I rather think I would have wanted to die.

All in all, this particular safety induction was pretty grim because it ended with the inevitable introduction to that piece of rope – the rope of no return – that was supposed to be thrown over the side of the platform so that I could climb down about eighty feet and then cast myself off, falling twenty feet into a tumultuous sea.

In times of distress most of us look upwards. Apart from the instinctive appeal to our Maker, offshore there was another more practical reason. In theory, thanks to the chopper, that was the direction that could bring our salvation. Apart from the questionable accessibility to helidecks, realistically we all knew that platforms are far from the base heliports. In truth, this sense of succour could be but a wishful dream.

My increasing journeys offshore made me long for a universal safety code throughout the North Sea. On one memorable occasion, I responded to an alarm with my usual commendable alacrity and with the great confidence of one who knew what he was doing. Then I discovered I had (a) gone to the wrong muster point; (b) not taken my life-jacket from my cabin; and (c) to cap everything, had misunderstood the alarm. I consoled myself with the view that you have to be quite clever to get things so wrong. But really I had no excuse. I had confused the safety system on this particular platform with the one on the previous day's installation. It was at such times that I longed for an offshore ecumenical movement which would encourage companies to have much more in common. And it was not only some of the alarm and safety systems that needed rationalising but the platforms themselves. The older ones, with all the bits and pieces locked on, were genuinely confusing. Newer ones had fewer of those interminable steel stairs. They seemed to have been constructed with a thought for the more simple-minded. Significantly, on the later designs I began to find my muster station.

'Safety' was a key word at all times and a deliberate flouting of the regulations met with no sympathy. There was a complete ban on the consumption of alcohol offshore. At one time there had been a very mild relaxation on 25 December and 1 January but the privilege had been abused by the foolish few, with potentially serious consequences. It was not dissimilar to those days when the Navy dispensed daily tots of rum and some matelots over-indulged in the 'sippers' and 'gulpers' of generously inclined friends.

I was going offshore a few days before Christmas and, as our bags were being searched at Sumburgh airport, I noticed that what can best be described as 'the heavy mob' were present in force behind the inspection table. They seemed to be particularly interested in two tins of coke that had been unpacked from a holdall. The owner was invited to leave them behind and accept a couple of replacements. This was declined and then the owner was invited to consume the contents there and then. At this point the prospective passenger expressed an urgent desire to return to Aberdeen on the next fixed-wing flight. It was all very ingenious. The coke had been extracted from the cans with a hypodermic syringe and had been replaced, by injection, with neat vodka. The holes in the base of the can were minute and barely visible. Obviously, someone had got wind of the plan and blown the whistle. Unlike so many cases, this was seen as entirely acceptable by the man's colleagues and there was little sympathy for someone who would be instantly dismissed, never to work offshore again. Smoking was allowed in clearly specified areas; parts of the PLQ, the Bear Pit and the Wendy House. Some of the most interesting discussions I have ever known took place in rooms so filled with smoke that it was a waste of tobacco to light your own pipe. Safety matches were supplied and no-one was allowed to leave shore with 'lighting material'. As I knew only too well, cigarette and pipe lighters could be impounded in Aberdeen to be collected on return – if you remembered! When I did remember, I would rummage around the basket full of surrendered lighters realising that mental aberration was not confined to the chaplain.

The attitude to safety could be very directly related to the human condition. The consequence of good morale and high self-esteem could result in a perceptible decrease in the accident level. Safety was not just to do with legislation and regulation but also with an understanding of the people at whom the decrees were directed.

One sunny day I was chatting to the OIM on a small gas platform off the Norfolk coast. The crane man, installed high above the deck area, was having a peaceful morning sitting in his cabin in quiet contemplation. All of a sudden the cab door opened and the operator swung himself onto the narrow ledge that surrounded the crane perched seventy feet above the sea. He was not wearing his safety harness but had armed himself with a cloth and a bottle of window cleaner. Slowly and meticulously he worked his way round his cab, spraying and polishing, blissfully untroubled by the sheer drop. The OIM was a wise man, 'If I shout, he will fall off,' so he held his peace and together we watched fascinated until the orgy of cleaning was over and the crane driver climbed back into his now gleaming cabin. (Thank goodness he did not step back and admire his handiwork.) It turned out that the operator had worked offshore for many years with an unblemished safety record. When interviewed, he was quite unable to explain why he had comprehensively broken so many safety regulations.

This incident was a lesson that served to underline the reality that when people are involved there will always be a degree of unpredictability in their behaviour. Surely that is one of the wonderful, albeit disconcerting, marks of our humanity? It is a sombre fact that no matter how much thought and effort goes into making offshore a safer working environment, it would be unrealistic to expect it to be 100 per cent incident free. Risk assessment is a fine art, with human unpredictability as the ever present wild card. When things do go wrong, bad luck and a chain of quite unpredictable events, with their domino effect, invariably play their part. Equally the unwritten history of the North Sea would be full of stories where good luck has played a part, and events that could have ended in disaster had a happy ending.

A driller regaled me with one story that made him, he said, dry in the mouth whenever he told it. By some extraordinary mischance he began to drill into a well that was already being worked by an adjoining platform. (The technical ability to drill laterally as well as vertically has a potential for problems.) The mathematical odds on this happening are enormous. Luckily (the only word) there was not much gas present. If there had been, the consequences would have been catastrophic. But for all the times that Lady Luck is on your side, there is the recognition that from time to time luck can run out and then a very different scenario evolves.

When delivering the address at a memorial service for those who had lost their lives in an offshore incident, I asked the rhetorical question, 'Just why should these tragedies take place?' and then sought to respond by saying:

> There is no easy answer, indeed I doubt if there is any satisfactory answer, and certainly not one that would make it possible to

abolish once and for all the ingredient of hazard that will always be present in any great enterprise. It is that very ingredient that provokes the challenge and demands the participatory response in a certain sort of man or woman. They are able to accept that, for all the training, for all the legislation, the committees, and the analysis, always there will be something that can go amiss. It is the realistic acceptance of that possibility, however remote it may be, that gives stature to all those employed in some way or another offshore. It is this acceptance that makes these people so special.

Experts are at work on the risk factors in most areas of human activity, but there is evidence that as our lives become increasingly cocooned, many people have difficulty in living with the reality that in simply existing there will always be an element of hazard, no matter how small. A nanny state can blind us to reality. We may cherish life but no-one can be immortal. Maybe that is to our advantage. Within most people there lurks some spirit of adventure and, in an increasingly protective society, many find their release in parachute and bungee jumping, extreme adventure holidays, sailing round the world and trekking in remote places. It is not unusual for responsible people to choose to be involved in something that demands a bit extra from themselves and in the process helps them to find out what they are made of.

There are job opportunities that have that inbuilt 'demand' factor where the possibility of the unexpected lurks to a greater or lesser degree. But, and there will always be a 'but', human fallibility and technical failure, together or separately, can produce catastrophe. Training, testing and development can reduce, but will never totally obliterate the hazard. It is that knowledge that adds the indefinable ingredient that gives spice to the activities of the more adventurous, and a cautious realism to those who work in situations where hazards can arise.

Most certainly those who work offshore would not like to be seen as the glamour boys and girls of the North Sea. They are simply people who are going about the business of earning their living in a specialised environment. It is this very ordinariness that highlights the hazards that can be met and which can remain invisible until the fateful moment.

One day a steward, who was off duty in his cabin, went to the en suite WC. He lit a ruminative cigarette (this was quite legal) and in an instant his world came, literally, tumbling around him. The toilet was demolished, leaving him very shocked and slightly scorched, maybe even more embarrassed! Unknown to him, gas had escaped into the toilet system. Investigation revealed that the corrosion of an internal valve had been the problem. Odourless gas, invisible and heavier than air, had been lying in wait – then Boom! It was this same gas that in another incident wormed its way down two floors in the PLQ and stealthily crept along at floor level infiltrating the galley. A chef bent down to light one of the stoves – fortunately his severe burns were confined to his ankles.

Being in a vigorous maritime situation it was inevitable that so many of the tragedies in which I was involved could be attributed to corrosion. Visible corrosion is one thing, and precautions and preventions can be implemented. The lethal danger is when the corrosion is invisible to the naked eye.

A worker stepped onto a steel cover to make a repair. The invisible supports had rusted away and the cover collapsed. He only fell a few feet, but in the process his safety helmet was dislodged and the crowded, jagged machinery brought about a fatality. The same fate befell an engineer removing a flange behind which a safety valve had corroded. A jet of water (under such pressure that it became like an iron bar) burst out and hit him full in the chest.

Metal fatigue was another invisible enemy. Because they were in constant use, cranes were regularly checked and serviced, but circumstances could dictate times when they worked to the limits of their tolerance levels. A collapsed boom, or even worse, a crane pulled overboard, was not unknown and were vivid reminders of the extremes demanded by some lifting operations which were within the prescribed safety limits but could take no account of unseen, quite unknown, structural weaknesses.

The drilling platform was the most potentially hazardous area of all. A drilling crew was always an impressively cohesive unit where everyone had a clearly defined task, and worked secure in the knowledge that everyone else was playing their part. The horror experienced by this tightly knit unit can barely be understood when they saw one of their team crushed by the crown block which fell when the securing hook cracked as a result of metal fatigue.

Sadly, such incidents have for long been recognised as the inherent hazards in heavy engineering, but there is one big difference. When such an incident takes place in an isolated location where living and working are irrevocably bound together, the effect of such tragedies can be huge. The fact that there are so few accidents attributed to man-made mistakes speaks volumes for the self-discipline of the workforce. I was always impressed by the acceptance of the dictum, drummed into everyone, that there was no room for human error. A constant state of safety awareness only worked because everyone recognised their interdependence and the need to support one another.

I retired in 1991 and seven years later returned to conduct the tenth anniversary services of the Piper Alpha disaster on the Saltire, Piper B and Claymore platforms – the latter I knew so well, I was made a VIP – an interesting experience – and flew out to the platform in a much smaller, sleeker and faster chopper than ever before. As I clambered on board I squeezed between the bulkhead and the pilot's seat. Unbeknown to me, I slightly tore my survival suit. On my arrival, the ever-observant helideck crew noticed the tear and, unknown to me, spent the whole day obtaining a replacement. It was a very personal reminder that a successful safety strategy is dependent on an infinite capacity for taking pains, together with an unceasing vigilance.

After all those years I understood even more clearly the fatigue which could be etched on so many faces. It was the visible reminder of the toll taken – not only in doing a good job, but in doing it safely, so that no-one's life was endangered. It is this awareness that has to be a part of you for every moment you spend offshore. I cannot think of any other form of human toil where, apart from the work itself with the caution it demands, people sleep 'over the shop' with a hazard potential and where commuting presents its own unique potential for peril.

The offshore industry has not escaped its share of incidents both great and small. The fact that there have not been more says much for safety regulations that are under constant review. But it says even more about those who are minded to observe them.

8

Tears and Fears

LOG EXCERPTS

⟡ *Held two services to mark the first anniversary of the Chinook disaster. Both in the cinema and the first one with standing room only. The second attended by two men, who came both obviously much moved.*

⟡ *This has been one of the most peculiar and demanding trips yet. The whole range of emotions has been covered and quite new situations have arisen. When a new trip begins you have no idea what may transpire. I realise that we have had as near to another 'Piper' situation as could be imagined. Listening to those involved talking about gas escapes 'sounding like banshees' is an exact replica of the comments made by Piper Alpha survivors. There is much for which to thank God.*

⟡ *The fatality was a very unpleasant, and public accident. The drilling crew in considerable shock, with the deck crew (very experienced, mature, sea-going men) doing the wretched business of clearing up. Talked with the drilling crew; so little that can be said for everyone recognised theirs can be a hazardous operation. But to be able to offer a listening ear seemed important. The drilling crew requested a service to be held in the cinema – packed out. Much time talking and listening and went to bed at 2130 but at 2230 two men wanted to talk. An hour later went back to my bunk – so ended a day that had begun at 0430 hours.*

'You are nothing but a lackey of the oil companies.'

Distraught parents of Piper Alpha victim

'I begin to believe there is such a creature as a second class widow.'

Wife bereaved in a 'one-off' incident

↻ *Spent a lot of time with the scaffolder's foreman and was taken to the place where he had died. I have done this too often and the fragility of human life is all too apparent.*

↻ *Up at 0600. Something made me have a shower immediately, and in the course of it all the lights failed (alarming!). There was then a red alert and everyone went to muster stations. Outside the wind was gusting 80 knots with very high seas. The FSU had broken from its mooring at the base plate and was drifting out of control. If the wind had been 20° to the North it would have produced a holocaust with the probable destruction of Fulmar and all on board.*

↻ *The morning spent in yet more listening and quite a large amount of discussion as to how one lives with a feeling of self-induced guilt. A watch should be made to ensure this does not develop into a real paranoia.*

↻ *Once again landed on the* Tharos *– still controlling the fires and keeping the platform (what remained of it) cool. Only three of us flew back to Dyce, one a remarkable American gentleman with black, embroidered, patent leather cowboy boots. Fighting fires and capping wellheads must be character building.*

↻ *For some of the men this is neither the first tragedy nor the first explosion, and there really are offshore veterans now who have been through a great deal on many occasions. I wonder if there could not be awarded a medal for service offshore. Campaign medals are awarded in time of war, but courage and determination are not solely confined to the battlefield.*

The first two decades of North Sea operations had not lacked incidents and consequent loss of life. But in the destruction of Piper Alpha in July 1988 came the defining moment for everyone. From the outset of offshore operations, it had been accepted that things could go wrong. There was accident potential in the constant journeying between the beach and the platforms, and the installations themselves were vulnerable in times of both exploration and extraction. But this was the price that might have to be paid for the black gold and the risk was accepted by an industry which was no stranger to potentially dangerous situations.

I was very much the incomer to this very special world. The one thing I could bring was human antennae which, over many years, had become finely tuned to patterns of human chemistry and one was aware of a presentiment, not often expressed, yet tangible, that a disaster of huge significance was bound to come. Those offshore, not least those whose work involved visiting many platforms, were shrewd and observant. They saw contrasting management styles and safety observance and in some offshore situations they were less than impressed. It was inevitable that they formed their own views where disaster might strike.

The trips offshore began in 1986 and, as I moved around, I became aware of this distinct unease amongst some of the most experienced workers. It was all a bit bizarre for there was nothing much to go on. It was to do with 'hunch' and instinct. But I did get the impression that some indefinable sands were running out. However, when the explosion occurred it stretched the credulity of even the most prescient. The total annihilation of a platform with 167 losing their lives and only 63 survivors far exceeded anything that could have been envisaged, and brought home in the starkest way the natural forces that had to be controlled and the demands that were made on men and machines.

At the time of the explosion I was out of the country, but was able to return within a day. The operating company, Occidental, felt that in the unfolding situation I should go out to the Claymore. This was Piper Alpha's sister platform and was located some twenty sea miles away. In the emergency, Claymore had been all but forgotten, partially explained by the severance of the communication systems – these had been designed to pass through Piper Alpha. With the extinction of that platform, Claymore was virtually incommunicado.

En route we touched down on the *Tharos*, the maintenance and salvage vessel which lay alongside the blazing remains. The platform's once massive structure was all but obliterated, with very little being identifiable. The scene was dominated by an ear- piercing hullabaloo coming from a row of fractured pipes that jutted out of the sea and were roaring away like a collection of gigantic bunsen burners. The pressure being generated was intense and the tangled metal was white hot. Powerful jets of water played on the wreckage in a vain attempt to lower the temperature and a curious mixture of steam and sticky smoke enveloped everything. The nightmarish picture was completed by foul air loaded with the acrid smell of fuel oil.

The few survivors had long since been flown to the beach and, looking at the scene, I found myself wondering how anyone had been saved. The refrigeration boxes standing on the *Tharos'* deck were a mute reminder that only the dead would now be retrieved. It was at this moment that I began to realise the concentration of sorrow and bitterness with which I would be involved for so long.

There were over a hundred men on board Claymore – not that you would have known it. Everything was shut down and the silence was total. Men stood in groups, barely exchanging a word, yet drawing support from the association. Others were trying to occupy themselves with some self-created task. If someone dropped a spanner the noise echoed through the structure. Even the usually restless sea seemed stilled, reduced to a glassy swell that slid up and down the platform's legs as though they were set in treacle. It was a windless, glorious, midsummer's day and in the distance a vertical plume of smoke reached up to the sky marking the spot where seventy thousand tons of steel had once stood.

As I moved around listening to the fury and the sorrow, I let it be known that I would be on the helideck at sunset and everyone was welcome to join me in a few prayers. When I climbed up there on that peaceful summer evening, it was crowded. The shadows were lengthening and still the black column climbed into the sort of blood- red sky that is only visible in a maritime sunset.

It was a dramatic backdrop as we said our prayers together, and I was acutely conscious of the inadequacy of words and the benison that is silence. After the Blessing I closed my eyes for a few moments and when I opened them my congregation had disappeared. No longer were a deeply saddened group of men standing with me. Instead, their silhouettes stood against the sky as a great silent ring of men encircled the helideck looking out to sea – each one alone with his thoughts.

Chaplains have conducted such services in all manner of circumstances ever since they accepted the responsibility for the 'care of souls'. Those who have no experience of such a situation cannot be blamed if they criticise the holding of such a service in this so called-secular age, for all manner of personal view-point is likely to be present. There will be those with a personal belief for whom this sort of worship can mean a lot. Equally there are those who wrestle with their uncertainty, or those who give no thought to things of the spirit, and for whom the whole exercise would seem meaningless. The reality is that, irrespective of the personal convictions represented, such gatherings are always found to be helpful. The reasons can be many and complex but there is little doubt that there is a strength drawn from a group united in a shared sorrow, with some of those present able, at a deeper level, to reach out to the loving God. For my own part I tried to articulate something of everyone's heartbreak – often the more stumbling the effort, the more helpful it could be. God can become so much more real in the shared pain of a traumatic experience.

Over the years of my chaplaincy such services were far too frequently held for those who lost their lives whilst working offshore, when workmates turned out in force. 'To pay your respects' is a bit of a cliché, but how else can you show your regrets and offer your condolences? Memorial services have a much needed therapeutic value, and the classic words of Christian comfort, for all the agnosticism present, seem to reach deep into the psyche and do some healing.

This was particularly noticeable at the many services held in the wake of the Piper Alpha explosion. Numerically, the most significant was the enormous memorial service held at St Nicholas Church in Aberdeen which spilled out into the church grounds, when it was estimated that some five thousand offshore workers took part. For me this was the day when the offshore family came into being. Never before, nor since, has there been such a concentration of the workforce and in my address I sought to express the reality of the offshore family to this huge gathering:

> Those who work offshore have their kith and kin on the beach, yet, as soon as they enter the heliport, they join their extended family. In the quite unique work situation of the North Sea, it is the very ingredients that make a home a happy home and a contented home, that have always been required offshore: good humour, honesty, integrity and quiet courage – nothing too complicated and yet very special.
>
> For let's be quite clear about the men who died. They were ordinary men, doing the ordinary jobs at which they were so skilled. It is the location that makes it all so extraordinary . . .
>
> This is a people industry and they are its most precious investment and give it its most special character.

It was significant that not only the employees of the company affected were present, but many others from other companies who felt just as involved. The spontaneous and universal testimony that 'we are all in this together' transcended corporate differences and bound the workforce irrevocably together.

The same signs of solidarity were shown in remote Highland villages and small Cornish communities where everyone would turn out in tribute to a son or husband whose work had taken him over the horizon. There were also the bewildered widows who spoke no English but discovered that heartfelt compassion was international and transcended differences in culture and language.

Out of the awfulness was born the united understanding of the hazards implicit in a common endeavour. Thus the offshore family was forged. Yet, how unfortunate that it took such a sombre event to bring everyone together. Perhaps in their combined wisdom the oil establishment could celebrate an anniversary that would make all involved rejoice in the huge tasks performed.

As for many others, the Piper Alpha disaster concentrated my mind. I was forced to review my professional experience, and assumptions, up to that date. I had to examine some of my unconsidered attitudes, and to revise others completely. Out of this time for enforced reflection, it became clear to me that in such incidents it was important to stand back until the situation resolved. Rushing into a critical situation and hoping to 'do good' or, even worse being anxious to be seen to be 'doing good' had to be resisted. All too often there can be a plethora of people who want to help but are not very clear how. They can clog up systems that have been carefully planned with tasks allocated to those best qualified to cope with them.

The role of the Christian Church has, for centuries, been one where it accepts responsibility for the *cura animarum* (the care of souls). This responsibility has never changed, and even in a self-styled secular society it is still very evident that people instinctively turn to the Church, or its representative, for the succour that at certain points in their lives can be so desperately needed. At the very least there are many who, having been bereaved, would want the deceased to receive the dignity offered by Christian burial, and for those agonised in their grief, and who are of a mind to receive it, there is always the message of everlasting life. For those who remain, family, friends, colleagues, and those dealing with the incident, the reassurance of a Christian love that can withstand the most desolating times has to be available. The learning curve was steep as I steeled myself to listen to the distress and absorb the accompanying anger. I did my best to acquire a sensitive empathy that could help me to understand how people felt – and then to translate these feelings into words. At the most practical level I tried to sharpen up my sense of awareness of the pastoral needs people might have. More generally, I did my best to take a lateral view that might uncover situations that had not been envisaged in the contingency planning. What became very clear was that no two incidents are remotely similar, and the creation of an umbrella blueprint that might hope to cover every contingency was unrealistic. In short, a chaplain was best utilised when his overview helped him to prioritise the areas where he had an appropriate contribution to offer. When you start to make such a list it can be surprising.

Always there are those who have survived. Over the years, crews on platforms become tightly knit units – it could hardly be otherwise when every alternate fortnight is spent together. When there is a tragedy the shock is profound. The empty bunk and the usual seat in the mess room bear silent witness to someone who will not return. The most harrowing tasks are carried through by those who have to restore everything to normal at the site of the accident. You cannot be cloaked in the impersonal detachment that can be acquired when you do not personally know the victim. To take one example, it is tough on the helicopter crew who have a long flight to the beach with a body bag containing a late colleague drowned in a chopper incident.

At the most practical level, religion can be useful irrespective of the strength of personal conviction, but I soon ceased to be surprised at the strength of beliefs sincerely held. There was something therapeutic in holding an act of worship on a platform for it provided a 'pool of stillness' in a very raucous world. It meant everyone came together. Always there was an order of service ensuring maximum participation. When necessary, the service would be repeated for the benefit of the night shift, or at any time that was helpful. On one occasion four men were working down one of the platform's legs and when they emerged and discovered they had missed the service they asked for a repeat. We sat around a table in the empty mess room and thought about our friend, his wife and family, and commended him to God's care – then they replaced their hard hats and returned to work.

When there was a fatality it seemed important to write to the next of kin and enclose a copy of the order of service and explain how we had remembered their loved one. At such poignant times the remoteness of a platform can be deeply felt. Many families had absolutely no idea where their loved one was located in the North Sea, and consequently there can be very little to act as an emotional focus. It was strange rarely to meet the next of kin but maybe this is how it should be, for the chaplain's remit was clear and was to do with events offshore. On the beach all the personal support services were available; so much so that there were times when those at the receiving end were in danger of being overwhelmed by a tidal wave of goodwill. For the bereaved it is wonderful when their specific needs are fully met, but there is a need to protect them from unwarranted intrusion. It's so easy to forget that peace and silence can have such wonderful healing properties.

The mid 1980s were remarkable for the number of major disasters in the United Kingdom. The Bradford Football Stadium had gone up in flames with a heavy loss of lives, soon to be followed by the Underground horror at King's Cross. If that was not enough, a packed Chinook helicopter crashed into the sea off Sumburgh – only two survived. Then came the explosion on

Piper Alpha, to be followed a few months later by the Boeing 747 blowing up over Scottish border town of Lockerbie. Prior to all this carnage there had been the air disaster at Kegworth bordering the M1 and the Clapham train derailment.

By the time this litany of distress was over, there were people who had come to believe that it was possible to be an authority in dealing with the human implications of a disaster and who seemed compelled to exercise their expertise at every given opportunity. Of course, it is possible to learn from experience that has been hard won through involvement in a tragedy, and it helps in being able to anticipate the needs that have to be met. But there can never be a universal panacea that will miraculously take care of all the human dilemmas that will emerge. No two disasters can ever be the same. Locations can range from the densely populated to the remote, and whilst the need for human compassion is a priority, it has to be employed discerningly. The more diverse the support available, the more effective would be the relationships established. As a clergyman, I had to learn the hard way, for the assumptions of a previous generation could no longer be held. Social workers too, had to learn the same lesson, for they could have a somewhat omniscient interpretation of their role within the community and their relationship to it. This was particularly true when their clients, until that particular moment, had had little, if any, experience of the ministrations of a social work department.

The emotionally troubled can be very vulnerable, and all too often I came across those who had been the reluctant recipients of counselling when all they had needed was a listening ear and empathetic understanding to help them through the bleakness of loss and overwhelming sorrow. There are the times when a loved one dies and when it is discovered that a trouble shared can be a trouble halved. However, this sort of sharing is very different to traumatic, unbearable bereavement, where it is necessary to seek a clinical intervention by a suitably qualified individual before morbidity becomes apparent and a pathological condition starts to develop.

I suspect my approach reveals a generation gap. As I grew up, I had had to accept the fact of death, and the reality that you can die young. Nowadays, death is not accepted easily, with most people feeling they have a right to life until they reach a great age. When that assumption is not realised there can be a strong resentment and the resulting distress is not understood, certainly not accepted, and mourning is not allowed to takes its natural course.

Frequently, I found myself wondering when an 'incident' became a 'disaster'. Prolonged links with the media led me to the reluctant conclusion that a minimum of five fatalities at one time was a prerequisite before their full attention was assured, with front page headlines and televised interviews. Always, with that sort of publicity, there was a knock-on effect; a wave of public sympathy would be generated, translated into spontaneous cash donations, which had to be responsibly dealt with through properly constituted relief funds, which could be a nightmare to operate. Our lives are rarely straightforward but the complexities of some are quite remarkable, with divided families and dependent children from unions not recognised in law. Partners could be as much in evidence as wives, and trying to determine a fair and just distribution of the public's benevolence could involve a lot of discussion.

There is much untidiness in all our lives. All those loose ends that we mean to sort out but which always seem to be left undone. When I gave talks on the lateral look at an incident, I would start by asking the rhetorical question, 'If you were to lose your life at this moment, is it in sufficient good order to bear total scrutiny?' Invariably there would be a significant silence followed by some nervous laughter. The 'clean underwear in case you are the victim of an accident' syndrome can be a timely reminder of the chaos of many lives which remains concealed until the unexpected happens. One of my most miserable experiences was to go and see a widow, only to discover that she had always believed her husband had ten days leave between each trip offshore. It was only when he was killed in a 'one-off' accident, that she realised she shared her husband with someone else. Now there were two women in great distress and so little that could be said that would make any sense.

As the years went by, it became increasingly obvious that financial considerations were obtruding into the grieving process. There seemed to be two reasons. The bereaved became more aware of the insurance implications in an incident and became more litigiously inclined. At the same time legal firms began to adopt the American system of 'no win, no fee'. In the midst of all the sorrow and tears, and there were a lot of them, there could be events that made you realise how shallow some relationships could be. Not every marriage was a harmonious union and, as in any area of life, there were situations where an endured tolerance best described the union.

Early one morning I received a call from the police who had been summoned to a 'domestic'. There had been a fatality offshore and the grieving family had gathered together and had become rather upset with one another. It was the considered opinion of the constabulary that, as the Oil Chaplain, the upset was more my concern than theirs. The family lived in a new development, so new that it had not even made it onto my map of the city, but I only had to stop the car to be guided by the noise (sound travels far at 3 a.m.). It emerged, eventually, that the upset was all to do with the anticipated insurance proceeds resulting from the death of the loved one. The family were split down the middle when it came to discussing the disbursement of the claim. Would they continue to live in the UK or would the unexpected windfall justify a move to Spain?

When greed is allied to anger, it makes a disturbing combination. To order flowers for the funeral of a loved one yet send the bill for payment to the oil company was, initially, hard to understand. But grieving next of kin could feel so powerless in these tragic circumstances, and they were angry. Inevitably the operating company was the whipping boy, and the suject of their rage.

There was a considerable difference between human reaction and media treatment when an individual fatality occurred in the course of someone's offshore work. There might be a brief mention in a local paper, but there would be no surge of practical sympathy from the general public; there would be no public appeal on behalf of next of kin and the financial settlement could vary enormously between the company and contract employees. This was disturbing and a Trust was established to provide immediate financial help to offshore workers' families who, in the event of a fatality, or a major upset prompted by the diagnosis of terminal illness, could find themselves in need of immediate support. Such situations would not be covered by public appeal. The greatest supporters of the Offshore Trust were the workers themselves who had been disturbed by some of the financial anomalies that had emerged after such incidents. Establishing such a fund was facilitated by the willingness of prominent men in the oil industry who were prepared to give of their expertise as Trustees.

I found the inequitable attention paid to those who were individual victims, in contrast to those who had lost their lives along with many others, to be hard to deal with. It was a vexing situation which I tried to cope with by visualising every victim of every incident, major or minor, as someone with their own home, loved ones, and dependents, and from this my personal formula evolved. Every individual, be they alive or dead, became a unit of concern where loving care, both spiritual and practical, had to be available. No distinction was to be made by the enormity of the incident.

My confusion came to a head when Piper Alpha exploded and a mystique developed which engulfed all that went on. By this time I had been deeply involved in a series of offshore tragedies and was very concerned to maintain

an emotional balance. I found myself involved in countless media interviews – there were few people the media could talk to who were not official
spokespersons for the companies involved. Whilst expressing from my
heart my sorrow and distress at a grim situation being endured by so many,
I was continually haunted by the recollection of a family who had known
similar grief when their loved one had died in equally tragic circumstances,
but because it involved an individual loss of life, there had been none of the
public attention with its inevitable result – an outpouring of emotion and
practical sympathy.

There is a brighter side to this sombre picture. In extreme situations,
there are always people who behave with humbling heroism – much of
which is largely unseen and will never be recognised. Many who have
worked offshore can tell tales of selfless action when an emergency has
arisen and an instant response has been required. The offshore saga is
crammed with examples of the best in unassuming support and so too on
the beach: those who count soiled banknotes and replace them with new
ones; jewellers who reshape and burnish wedding and signet rings; watch-
makers who clean and repair timepieces. All this, so that next of kin may
receive their loved ones' most personal possessions in pristine condition.

Most people are on 'a high' when they respond to an emergency, but
the longer a critical situation continues, the more the capacity to be noble
begins to sag. In prolonged incidents – such as the aftermath of Piper Alpha
– as the days dragged by, there was a decline in personal dignity. Anger
(perfectly understandable) and greed (not so understandable) when put
together are unpleasant enough, but when a bit of self-righteousness is
then added the mix becomes very potent.

In a major crisis the media feeds the egos of self-seeking publicists, and
self-appointed experts were given press space and air time to express their
entirely subjective views. Piper Alpha saw an abundance of this behaviour
and as the days and weeks went on, it began to infect the relationships of
those whose sole purpose was to support the mourners and survivors. There
was no united front, and people who should have known better, amongst
whom I included myself, ceased to communicate with one another. I made
the mistake of assuming a pastoral role that many, through conscience, were
unable to recognise. I should have had greater confidence in the ability of local
communities to support and succour the distressed in their midst. I need not
have worried because they did the job anyway and my misguided view was,
rightly, ignored. It was tough to accept that there was no need for conflict; that
the different approaches to the bereaved were complementary, and offered

a very necessary choice. My steep learning curve continued when I met the spokesman for those who had been most directly affected by events, and who had become a rallying point for others equally distressed. Intriguingly, some of these individuals did not have a direct personal connection with the crisis. But they all had one thing in common, a personal dynamic whose reasons were not identified at the time; even by themselves. It was fuelled by one or more of many things: anger, guilt, political bias, social sensitivity, a sense of personal failure, a long held grudge. These are the sort of instincts that can work within us all at one time or another and can energise us in a particular direction. Such a dynamic can make it uncomfortable for those whose own agenda is less complex, and more basic in its compassion.

Over the years, several memorial services were held in the Kirk of St Nicholas in Aberdeen, and inevitably, in the eyes of the offshore family, it came to be associated with sad occasions. This perception was revised thanks to the generosity of the oil industry. It was possible to build a chapel within the north transept and, when it was completed, it proved to be a wonderful example of contemporary Scottish craftsmanship. It is dominated by a stained glass window, commissioned to portray all the main features associated with the offshore oil industry. Amazingly there will come the day when it will be the only visible reminder of an extraordinary page in industrial history. Apart from being a holy place for prayer and meditation, the chapel has come to be much used for weddings and baptisms, which goes some way to restore balance to a building so closely associated with sombre occasions.

Marking the anniversary of a tragedy can be a delicate area. There are those who will wish to remember, not least the next of kin. Conversely, survivors and colleagues find it to be a time when the wish is to look forward and not back. Inevitably, the first anniversary of the loss of Piper Alpha was particularly complex. It was well attended, but the most poignant moments occurred after the onshore service was concluded and the families of those who had no known grave embarked in the SS *Sunnivar* (known to all who have sailed from Aberdeen to the Shetland Isles). She arrived in the early evening at the marker buoy where the huge platform had once stood. The weather was remarkable, with a sea so benign that there was not the slightest suggestion of a swell, and a sun so hot that, on board a vessel that had tried to anticipate every need, practical and emotional, the one item most required had been omitted – sun cream.

The sea was so calm, the air so still, that when the ship's engines shut down all on board were enveloped in a vast web of silence. A cascade of floral tributes was released into the sea, floating gently with the current over the ill-fated site. The only sound came from those who began to cry very softly as they remembered their loved ones. The poignancy of the moment was unforgettable.

So often I found myself wrestling with the perennial question: did the Chaplaincy serve a useful purpose in times of tears? Certainly not to everyone. On one platform after an incident, someone was moved to write to the official publication of Cruse, the society that offers support to the bereaved, suggesting that my platform visits were useless. It was disconcerting news and hard to take. I tried to follow it up and learn how my contribution could have been more significant. All efforts were in vain and to this day I am still ignorant of the reasons for this outburst. But it served as a timely reminder that no matter how hard you might try, and how sensitive you sought to be, there were always those who saw you as an intruder in the closed world where, for some, there was no room for belief of any sort.

Increasingly, safety became the paramount concern of a shaken industry and that must have been for the general good. In the immediate aftermath of Piper Alpha I was interviewed by Health and Safety officials – disconcertingly they seemed more concerned with the revision of rules and regulations and less with the life style and rhythms of work that can help to make people safer. A real plus was the realisation by the offshore family that it had a very real identity of its own. People were noticeably more reflective and supportive of one another. The importance of their work had been affirmed and, whilst they had to live with their sorrow, they got on with their job.

The tears shed offshore, of which there were many, were not an indication of weakness. Rather, they showed the corporate strength of a body of men and women who lived out their working days within the context of an unspoken, but real, tension – which in a rare and unsuspected moment could be justified. They accepted the possibility of calamity as an integral part of the job they had chosen to do of their own free will.

Enroute in a Sikorsky 61, the workhorse of the North Sea. (p. 11)

Stores asssistants have no clout. (p. 17)

Working alongside
Geoffrey Shaw.
(p. 18)

A bus courier with
Scotia Tours. (p. 19)

Wee Blondie's first and my
last offshore trip. (p. 32)

Independently inclined, well motivated and
resourceful. (p. 39)

Urban man becomes
maritime man. (p. 40)

The legs of a platform can be the
height of St Paul's Cathedral. (p. 49)

Much work is carried out in isolation. (p. 49)

The experienced chefs who had cooked their way round the world. (p. 108)

A member of the crew presents the Princess Royal with a large cheque. (p. 50)

Christmas offshore was show time for the galley. (p. 57)

On board Statfjord B with the chaplain (Terje Bjerkolt) and OIM (Jan Larsen). (p. 57)

There is no such creature as a typical offshore worker. (p. 36)

The feel of a united family. (p. 59)

There was something
therapeutic in holding
an act of worship. (p. 92)

Floral tributes. (p. 98)

Like a piece of rusting meccano. (p. 100)

Left: The control room with its warmth and soft lighting. (p. 101); *Below:* A descent to sea level. (p. 103)

The chapel – a very special spiritual base. (p. 120)

Norwegian offshore chaplains – all of them! (p. 132)

Life offshore does not encourage sartorial elegance. (p. 136)

The Moderator went
his rounds. (p. 106)

An activity for the
younger man. (p.145)

9

Times and Seasons

LOG EXCERPTS

All day the fog has been slowly gathering, and by night time thickly enveloped the platform. The reflection from the flare was quite Wagnerian and its noise bounces back and can be heard in the cabin along with the foghorn. There has been no flying since lunchtime therefore no crew changes and some resigned frustration. Apparently the fog could last for three days!

'The best time in my life'
Stewardess offshore on
Christmas Day

Said the final Christmas Dinner Grace – the fifth! – at 0130hrs on Boxing Day. The night shift had come on duty at 1800hrs on the 25th.

Men want to come and talk. Five such visits on Christmas Day to my cabin. Very stormy with high seas and winds over sixty knots. Very much the time to think of the crews of the supply ships and stand by vessels. At the crew's request I held a service at midnight in the quiet room. About 30 persons attended and the doctor (with one finger) played the carols on the electric organ – they were well sung and it was obvious that those who came were glad they were there. The OIM (to the surprise of some of the crew) – read the Christmas Story.

The Christmas games programme had moved to its climax and I presented the prizes at some grisly hour in the morning. That is one of the features about life on a platform – there is really neither day nor night apart from the work patterns. It is also exacerbated by the latitude, with extreme darkness in winter and unending daylight in summer.

The live TV broadcast for Christmas Eve went off reasonably well. The singing would have been better another semi-tone higher! After the service, we all had a glass of (non alcoholic) punch – we were now in festive mood and everyone adjourned for supper (!). To bed on Christmas morning at 0230hrs.

Weather awful. Miserable flight in the chopper to Sumburgh. Fixed-wing onward to Aberdeen was no better. My neighbour clutched a 'comfort bag' throughout the flight. Jennifer met me at Aberdeen. She said we looked a particularly dishevelled group of passengers – too right!

It is time to put the record straight because it might seem that North Sea oil operations are unremittingly grim with little light and even less laughter. That is far from the case and it would be absurd to suggest that all the time you are offshore you are surrounded by grim grey faces etched with fatigue. There is much laughter and much humour. It is important to remember that for all the bad moments when commuting, the offshore worker does not endure anything like the daily discomfort of the regular rush hour traveller on the Bakerloo line. At the heliport it was to be expected that people could be subdued as they started out on another trip. After all, starting another week's work for anyone can produce the same sensation. But this Monday morning feeling could soon disappear if the sun was shining and the delays minimal. A good flight, fine bright weather, and a gentle breeze wafting over the helideck, gave even the most insensitive the chance to appreciate something of the very distinctive beauty to be found in the offshore world. The blueness of the sky and sea would contrast with the reddish brown of the platform's Meccano-like superstructure. Often, when I looked at it, my imagination got to work and I saw it as a span of the Forth Rail Bridge that had been crushed into a cube. Far below the helideck the sea would be lazily rising and falling around the platform's legs and the crane, usually a bright yellow for maximum visibility, would be fussing away, providing a vivid splash of mobile colour. Perhaps this is all rather poetic, but I found there was, visually, so much that gave my sense of the beautiful a big boost.

At night the senses could be further stirred. The arc lights, as for any stage, would heighten the sense of the dramatic. Deck crews moved in their well drilled routines, and the movement of materials was controlled by the deck foreman with the authority and hand signals worthy of Simon Rattle. And, all the time, there was the background music of a roaring flare, working machinery with the thump thump of the piping as the unending drilling went ahead. Not least there was the sound of the sea which could vary from a gentle hiss to a violence that could make the platform shudder. The cast were clad in costumes – dramatic in yellow and orange and designed to repel any storm. The scene was made even more memorable by the smoothness of the teamwork, and the blissful lack of awareness of the participants in their skills.

The seasons made their own particular contribution to the rhythm of life offshore. At 60° North, winter is marked by the wildness of the weather and the few hours of daylight. On the platform no-one ventured out of doors unless it was necessary. Drillers would disappear into their wigwam-like structure that shrouded the drilling floor and gave some protection from the elements. Deck crews were fortunate if their precious space was packed with material. Crates and drilling pipes could act as some sort of wind-break. The rest of the crew would disappear into the hanger-like structures which enclosed the extraction machinery. The technocrats were yet more fortunate. They would amble off to the control room with its warmth and soft lighting. It was not unknown for someone to get through a whole trip without once emerging into the fresh air. This was in vivid contrast to the helideck crew who never lacked copious draughts of sea breeze. They would dress to look like astronauts and one would man the foam gun, always directed at the chopper, whilst the others supervised the arrivals and departures and shepherded everyone clear of the treacherous rotor blades.

On the beach there is a general perception that everyone, all the time, has to be kitted out as though they are sailing round the Horn. Surprisingly, once you had climbed down from the helideck, there were few points from which you could glimpse the sea. The steel ladder with its gratings con-necting the different levels could provide awe-inspiring glimpses of what is best described as 'marine turbulence'. There could be little shelter, and in winter the chill factor was a forcible reminder that the wind blew directly from the Arctic. I was ever mindful of those who sailed in the wartime Arctic convoys; of the deep-sea fishermen who endure the unrelenting conditions as a matter of course; and of the bobbing corks that were the ever-vigilant stand-by vessels. It seemed to me that there was the common strand of belonging to an island race. However urban we may be, there is still a response to the call of the sea and a certain nonchalance in facing its rigour. I believe this to be true but there remained that one big difference. On the platforms we were immobile and stuck in one place.

Summer could not have been more different. There were the months with long, long, days and once the daily flights were over, the helideck became the haven of the sun worshipper. Their work done, large numbers of the crew would sprawl in their unbuttoned ease, surrounded by the enthusiasts pounding out their circuits, as they tried to remain fit for some charity marathon, mini or maxi, on the beach.

With the better weather, flights were more likely to run as scheduled but winds could make a mockery of timetables. Running into a gale could be like making progress through glue; conversely a tail wind could bustle a chopper along in a most unladylike manner. And yet I always remembered the throw-away remark of one pilot who 'preferred a bit of wind'. Seemingly it gave the chopper something to fly against when approaching the helideck.

There were summer days when the journey started idyllically and then suddenly everything could change. A North Sea haar would descend. It could be very cold, and very impenetrable, and seemingly very immovable. There were times when you began to believe it had become a permanent fixture. Schedules would be revised and a whole day could be spent at Aberdeen airport admiring the runway basking in the sunshine whilst the Shetland Isles were totally closed in. Once again a hot sun and a very cold North Sea had started a meteorological tussle that claimed its innocent victims.

You could take your choice. Whether you preferred the clinging, all enveloping, summer fogs or the crashing seas, roaring winds and long winter darkness. The contrast was complete – suffocating silence or an elemental cacophony of sound; I hated them both.

Weather conditions that seemed idyllic could produce their surprises. From time to time I would receive requests from widows of men who, at one time, had worked offshore, asking me if I would scatter their late husband's ashes at sea. The first such request came early in my chaplaincy when I was particularly ignorant.

I took the casket offshore through a bemused airport security. Offshore it was a gorgeous summer evening with lots of sunshine and a mere zephyr of a breeze. I asked the crew of the platform if some of them would make up a congregation and many were kind enough to oblige – no doubt curiosity gave some impetus to the numbers. I gave a thumb-nail sketch of the deceased to the assembled company and then we proceeded to the Service of Committal. At the appropriate time I took the casket to the rail, and in the stillness of a mid-summer evening, emptied the ashes carefully into the sea far below. How stupid can one be? When you are eighty feet above the ocean you forget about the draught that flows in the calmest conditions beneath the structure. In an instant everyone was wiping their eyes – not with emotion – brushing down their dungarees and gingerly replacing their hard hats. So much for the dignity I had sought!

The story does not end there. Four years later, on a dark stormy night, I was conducting a similar service, but on another platform. The usual faithful congregation had gathered under the arc lights. At the appropriate time the casket, complete with heavy shackle, was belayed into the sea. Everything went well and with due decorum. After the Blessing, I thanked the small group for their presence. There was a slight pause and then a voice piped up, 'Well Andy, let's face it, you did a better job than on the Dunlin!'

I suspect that my now notorious service had done the rounds of the North Sea – as did all the gossip – for on another occasion the OIM announced that as the weather was well nigh perfect, with little swell, he proposed that I descend to sea level in a lifeboat and conduct the service as close to the ocean as possible. There was a positive side to this committal, for a brief excursion in the craft round the platform let me take in, at sea level, the enormous size of the platform – hard to gauge from any other angle.

Christmas offshore was a curious time. It seemed to emphasise a platform's isolation. To a degree that was because very few people really wanted to be there, and however hard you tried, thoughts of home and loved ones were never far away. Officially, too, it was cut-off time when official communications to and from the outside world were kept to the minimum. On Christmas Day, unless there was an emergency, there were no flights, no papers and the skeleton staff in the offices in Aberdeen were not inclined to make phone calls.

Sometimes there was an attempt to introduce some festive cheer (goodwill was never in short supply) but for many it was a time of mixed emotions, and for many their feelings found an outlet at the Christmas Eve Service. Always, there were those who had chosen to spend the festive time offshore. The agony of a marriage break-up is profound at any time but all the associations with the 25 December could exacerbate the misery and it was particularly traumatic for those with children. Perhaps getting away from it all was the best solution but it was emotionally costly for everyone involved. There is something abroad in the air at that time of year which crumples personal defences and prompts the exchange of confidences. Without exception it was the time when I learned about the anxieties that weighed upon my fellow crew members. They covered the whole gambit of emotional involvement and personal concern.

One expressed anxiety, albeit less disturbing, had me heading to a particular address as soon as I returned to the beach. A member of the crew had recently married a girl from the Philippines who had been a school teacher in her homeland. Her trip to Aberdeen was the first she had ever made outside her own country. Her husband was a kindly and thoughtful man and very aware that his wife was not used to northern climes. Before his departure offshore he had turned up the central heating to its maximum in their new, double-glazed, draught free home. Unfortunately, he had not shown his beloved how it could be controlled, and when I arrived at the house it was to the warmest of welcomes that modern domestic technology could devise. The poor girl was living in sauna-like conditions for twenty-four hours a day. The newly-wed knew no-one, and although an excellent English speaker, was shy in making contact with her neighbours. The only way she had been able to cope with the jungle-like conditions was to go for long, lonely, walks in an Aberdeenshire winter. Together we explored the cellar, found the heating controls, and moved them from 'max' to 'min'. Mercifully, the tropical outpost in North East Scotland soon disappeared.

What did not go away were the difficulties that could be faced, not least by the newest of offshore wives, in a relentless life where you had to reconcile yourself to sharing your man, no matter the season of the year, with a North Sea platform. Problems could be further aggravated because offshore families could be so scattered. An understanding, knowledgeable, community support was not always readily available in times of need. This was in vivid contrast to the experience of so many workers in other industries who could all come from the same area and be engaged in the same tasks.

Wherever you travelled, quizzes abounded, and on some platforms were taken particularly seriously by the contestants. Usually my answers were so dismal that my fitness to be able to do anything was seriously, and audibly, questioned. That was the downside, but the upside from an early exit from most tournaments meant that I was able to appreciate the encyclopaedic knowledge on display: it reduced Trivial Pursuits to a game for tiny tots. I was never able to work out why the galley staff, on every platform, were the masters of the art of the quiz. Their capacity for miscellaneous information appeared infinite.

The general awareness of people on the beach about those who worked offshore could be disappointing. However lamentable, I could sympathise, for it had not been all that long since my own life had proceeded blithely on its way with never a thought for the offshore family. Offshore activity could seem so far away and was easily forgotten. Maybe it was not unreasonable to hope that in church services when we prayed for others, some congregations might have had their memories jogged for those over the horizon. Yet, in all the worship I attended as a member of the congregation, I never once heard mention of the offshore worker. But there could be pleasant surprises.

The head of a Northumberland primary school wrote and asked for details of life offshore. He wanted to devote his Christmas Service to reminding the children about all those who would be away from their homes and working while they enjoyed all the festivities with their parents. On another occasion I opened an envelope and out fell a cassette. It was a tape from a group of children singing a carol they had written at school for those offshore. I took it with me and played it on Christmas Eve – there was not a dry eye in the house! It was like lighting an emotional touch-paper and the feelings ignited were a vivid reminder of the indefinable tension that was always there.

Visitors offshore are something of a rarity and that was small wonder, with seats on choppers at a premium and a price. There was, however, a steady trickle of leaders in the armed forces and politicians, local and national, making the journey to see at first-hand what was going on. The day was usually spent in familiarisation, with the mysteries of the drilling floor and the 'christmas tree' being revealed in all their glory. Once, my arrival on a platform coincided with such a visit. The VIPs were gathered in the departure lounge, suited up and ready to return to the beach. I asked them for their impression of the accommodation and eating arrangements only to discover that they had not been shown them. Maybe I should not have been surprised because there could be a very evident mind-set offshore where technology and production reigned supreme to the exclusion, once again, of the soft side and the more human considerations. But things can, and do, change.

It became apparent during my time that more thought was being put into colour schemes and living quarter layouts. Gone were the times when everything was set at right angles. Interior design began to be taken seriously. One day I had the pleasure of meeting such a specialist. She had just been contracted to renovate the living quarters and was spending her first trip offshore examining possibilities.

There came a day on one platform when, in an inspired moment, it was decided that every cabin should get an easy chair. They were duly transported across the North Sea and once the container was hoisted up and opened there was revealed a load of comfortable looking furniture. But there was a snag – a big one. It was discovered that no amount of ingenuity could manoeuvre them through the cabin doorways. It is easy to lampoon this sort of mistake in normal circumstances, but the situation becomes even more awful when existing furnishings are permanent and immovable. No amount of the twisting and turning beloved by professional removers could get them through the cabin doors and high hopes were spectacularly dashed. This lack of awareness was in evidence when seeking to respond to women's requirements which were modest, but different. Shelving and hanging space appeared to be demands that were not easy to meet.

From time to time, I came across two schools of thought which were in frequent collision. The first represented the historical understanding of the oil industry as a rough, tough, macho world that was here to stay. It was the traditional 'Aye been' attitude that is met in all institutions. The view was typified by the belief that within the industry all that was necessary was good food, a fair wage, and a clean bed. With these provisions the job would be done. This point of view was illustrated one evening, at a very grand dinner, when a senior oil executive made that precise point to me. He was amiable but very frank – a lethal combination – emphasised by eyes devoid of any hint of humour. He was one of the group of oil professionals who found it very hard to accept the concept of a chaplain concerned with 'soft' issues as being a legitimate part of oil situations.

The second, and for me, more acceptable, point of view was found amongst equally senior men who were acutely sensitive towards the people who were devoting their lives to making the offshore enterprise work. In their view the workforce, doing its job safely, deserved nothing but the best and they sought to provide it.

In May of each year, the General Assembly of the Church of Scotland meets for a week in Edinburgh. To all intents and purposes it is the, somewhat prolonged, AGM of the Kirk. Committees report and recommendations are debated. Presiding over the proceedings sits the Moderator (Chairman) who is a prominent ecclesiastic elected by his peers. He (and it was always a man in those days) holds office for one year. On several occasions it was possible to arrange for a moderatorial visit offshore. It was not easy for these senior clerics trying to adjust to this very different world. Platforms rarely receive such visitors and some of the encounters when the Moderator went his rounds, were thought-provoking. The hard hat and boiler suit are great levellers (rather like pyjamas in hospital) and as the churchmen were shown around the platforms they were the centre of much discussion. Amongst the workforce inhibition is conspicuously absent and the frankness of question and expression of point of view could have been disconcerting to lesser men devoid of humour. Prior to the official visit I would go out to the host platform and try to paint a verbal portrait of the office of Moderator. No matter what I said, the generally held view was that he was my boss – 'Just checking you out Andy.' My incoherent explanation of the Presbyterian understanding of the parity of the ministry met with complete incredulity. 'You mean you have no-one to answer to?', 'I thought a bishop, or a cardinal, was your gaffer?' The idea of a committee supervising my activities seemed to them even more incredible and, it has to be said, to myself as well. Every instinct prompted me to make the smooth response and say I answered to God. But I had a strong feeling that oil production might have been impaired whilst that one was being sorted out.

The ignorance concerning Scotland's National Church was an eye-opener. Even more disconcerting was how little it mattered to the majority of people. Conversely, there was a great interest in the eternal verities. Goodness, truth, evil, falsehood and not least life and death, were the source of endless discussion and debate. Probably it has been ever thus. Yet more intriguing was the total lack of concern in those denominational issues that so preoccupy the institutional Christian. There was a strange amalgam of disinterest and intense curiosity – probably best summed up in the experience of one offshore worker. He was travelling south in a train, when, at Stirling station, his compartment was invaded by what he described as 'a Dick Turpin like figure' clad in a black cloak and sporting a tricorn hat. The Moderator, for such was this splendid figure, then proceeded to quiz my friend about his church-going. For him it was a memorable experience that did nothing to create an understanding of the Church: it made it more far-fetched and unreal.

As well as being sanctuaries for some of the human race, by their very isolation, the platforms could act as resting places for many birds who had embarked on their long migratory flights and were blown off course. 'Twitchers' could be found on most platforms. They may not have been the greatest ornithologists but they did their best to feed and water their feathered friends. It could be a sad business because once the birds felt sufficiently rested they sought to be on their way. You could watch them take off from the platform and all too soon their strength would ebb away, they would lose height and then disappear beneath the waves. I always found this scenario to be very moving. Watching the birds fly away to their doom was frustrating for there was nothing one could do. Often I was reminded of the human parallel where the headstrong allowed their optimistic aspirations to override their ability to fulfil them.

There were opportunities to keep fit – not taken up by everyone – but there were enough enthusiasts to underline the inadequacy of the ventilation in the small gym. Also, from the adjoining sauna there would emerge a steady procession of human beetroots. Not least there was snooker, always snooker, but for a lot of the time the game was incidental to the chat going on around the table. As I watched, I acquired a wealth of miscellaneous knowledge on subjects as diverse as the traumas experienced by the South Shields Pool Association to the inner workings of the Scottish prison system. Snooker was just one part of a trinity of modestly energetic pursuits – the others were darts and table-tennis. But it is an indication of the splintered nature of the workforce and the universally held feeling that once you returned to the beach you left your work, and the life surrounding it, behind you, that there was never a North Sea Championship. The one group who worked offshore but who got together on the beach were the New Zealanders. They were worthy representatives of their tough nation, and inevitably, had formed a rugby team. They were hardy and fit and on the rugby field constituted quite a threat. The *coup de grâce* must have been when their opponents discovered that the Kiwis called themselves the 'Oil Blacks'.

The experienced chefs, most of whom had cooked their way around the world, many times, on liners and cargo vessels, were very alert to sensible diets and did their best to offer splendid examples of what some over-optimistic menus on the beach would describe as 'good, wholesome fare' – but could never quite deliver. The platform galleys deliver. Always they were ready to make celebrations special with anniversary cakes and they were very accommodating when it came to personal food favourites. Quite early on in my chaplaincy I must have publicly confessed to a weakness for bread and butter pudding (with custard). Thereafter, if there was forewarning of my arrival on a platform I could be certain that B and B pud was on the menu. A personal problem could present itself if I visited two, or more, platforms on the same day! There were rare occasions when the food supplied failed to reach the usual high standard and the dip in morale was noticeable. But which came first? The 'cause and effect' debate could be intriguing.

For all the long hours of work, life offshore was more than working, eating and sleeping with a small amount of recreation. These were not self-centred communities, but ones that were very conscious of the needs of a world that could seem so far away. The funds raised for good causes, which were very much a part of the human structure on every platform, were rarely publicly reported, but a number of hospitals are grateful for the provision of some long-sought piece of specialised equipment. Many a sightless person confidently walks the streets thanks to an offshore-sponsored guide dog, and there are lots of patients whose kidney failure had been ameliorated by a dialysis machine. It was not unusual for individual cases of need to be brought to the attention of the platform charity committee; perhaps a special bed for someone permanently immobile or the latest in electric wheel-chairs. Requests brought to the attention of the committee by a crew member, aware of a particular need in their own community, were invariably sympathetically considered.

The introduction of satellite television onto the platforms was a mixed blessing. Certainly it transformed the life-style of everyone but, it seemed to me, at quite a cost. The arrival of videos showing all important football matches had been a feature of life on board with an evening projection that brought everyone together as noisy and enthusiastic spectators. This all came to an end with the installation of TV sets in the cabins. No longer were there the rowdy, fun filled evenings; they were replaced by everyone retreating to their cabins to watch the big games behind closed doors. The social welding provided by communal tele watching came to an end and one of the invaluable ingredients in making a crew a cohesive unit disappeared. This was sad, for many crew members remarked to me how they had discovered the reality of community which compared so favourably with the life style they knew on shore.

You cannot miss what you have not known, and usually it was the older members of the crew who would wax nostalgic and lament the social disintegration. Others were of the view that it had no place in the maritime situation. I personally became less enamoured with a situation where the mess room became increasingly deserted and the opportunities for those chance encounters, which so often turned out to be mutually fruitful, began to disappear. It was these accidental meetings that could be so meaningful and in some ways were the main purpose of any offshore visit. The social anthropologists could have a field day researching the impact of television on small isolated communities, and a North Sea platform reflects this change in microcosm.

Although advances in technology began to destroy the sense of community, there still remained the old-fashioned, stimulating and provocative benefits of group discussion. There seemed to be something about offshore life that prompted a reflective spirit – or could it be the other way around? Anyway, there was great discussion, usually in the sometimes inappropriately named 'quiet room'. Apart from debates on sport, religion and politics, and among Scots – to the bewilderment of other nations present – there were times when these topics were not easily divided. I found that my mere presence would trigger stimulating talk about deep issues: the reality of evil, the need for forgiveness, the nature of true love, the purpose of life, and, inevitably, the place of religion; subjects spontaneously emerging and the source of vigorous debate. The tabloids often provided reports and articles on contemporary issues that were opinionated and provided a good base on which to earth discussion. It was a lesson for me. If a profound theme had been posted up as the topic for discussion I doubt it there would have been many takers. But when the subject emerged from a crumpled edition of the *Mail* or the *Sun*, then the conversation flowed, with such vigour that others who had been keeping their distance would drift over and join in.

I took part in more realistic discussions on great themes in a few years offshore than I had known in the whole of my professional life. Of course the location had much to do with their success. There would be no interruption, and there was no problem in getting home. The place was right, and probably the time too, for there can be an openmindedness that comes with relaxation after a hard day's work – and a good meal.

I learned much, not just from sincerely held views, but the picture that emerged was of the inflexibility of so much institutional religion, which can sit in its ecclesiastical fortresses instead of emerging to play away matches where it would be in a minority of one. The vulnerability of the clergy (me!) in these situations did much to put everyone at ease and no-one felt threatened or 'being got at'. One evening in the middle of a debate a voice remarked, 'It's not every day that we have a captive clergyman at our mercy!' Increasingly I felt that that was precisely how it should be.

There were invaluable reminders. Because someone might not be very articulate did not mean their views were second-rate and not worth hearing. Life (with a capital 'L') is an infinitely greater university than anything offered by academe, and its graduates have much to give. Glibness appears to be one of the marks of modern men and women, and to be articulate and have the ability to make good 'presentations' are deemed to be important parts of the 'success' package. But in the discussions, as often we would struggle together to describe our inmost feelings, I would be humbled by those who had come through one rugged experience after another and yet, remarkably, remained intact with an unshaken integrity and hope.

Often I reflected on the instinctive assumptions which those who have had a privileged and prolonged education make about themselves and their fellow men. At my ripe old age, late fifties, it dawned on this chronic late developer that of supreme importance is not how much we know, but waking up to the fact that we know so little. What a wise, but unknown, man he was who wrote, 'True wisdom can only be achieved when you are able to accept the poverty of your knowledge!'

10

Families and Friends

'Each time I hear on the radio the words North Sea, my heart misses a beat.'

Wife of an offshore worker

LOG EXCERPTS

⌗ Made arrangements to meet with one crew member on the beach, to help try to disentangle some of the complexities of his personal life. This turned out to be an ongoing activity for several months.

⌗ An interesting discussion on marriage with the helideck crew in their wee howf. You could almost hear his colleagues' jaws drop to the deck with a thud when 'Big Sandy' told us how he had spent a fortnight in the Philippines meeting possible partners.

⌗ Had a particularly long talk with a member of the crew whose daughter was in her second year at university and a son who is in Craiginches prison in Aberdeen. He (the father) is genuinely bewildered by the different routes taken by his children. He maintains that everything possible has been done for his son, but it does sound like the classic case of a younger brother overshadowed by a very able and attractive sister.

⌗ Was airborne by 8.30 a.m. and arrived back in Aberdeen in time to attend the Lord Provost's Appeal Committee. The sartorial elegance of those present went into sharp decline with my arrival.

⌗ Carrying a casket containing ashes is not the easiest object to satisfy Aberdeen security. But when I suggested opening up everything there *was a yielding of their position. At Sumburgh an astounded young security officer handled the casket as though it were the crown jewels – to the ever increasing curiosity of the rest of the passengers.*

I had hardly climbed into the taxi before the driver began a story he obviously could not keep to himself.

'On this particular day there was a lot of news on the radio about "an incident in the North Sea" and my passenger was listening intently as I was driving him home from the heliport. When we got there he leapt out and ran up the garden path whilst I got his bags. I looked up from the boot just as the front door was flung open. When the woman inside caught sight of her husband she promptly fainted.'

This little cameo from a taxi driver's day highlights two significant facts about offshore life. Many women whose menfolk work offshore have a well concealed yet gnawing unease when they are away on a trip. This is heightened because so often they have not the faintest idea where their man may be working. The North Sea covers a vast area and if the worker is contracted to carry out specialist work he can be very unsure about his ultimate destination.

It can happen so easily. When a worker checks in at the heliport the name of the rig or platform is displayed, but there is no indication of its location. Not very surprising; for just how do you give a meaningful description to a small speck in the middle of the North Sea? Even a rough description; something like 150 miles north east of the Shetland Isles or 120 miles east of Dundee might help. In fairness though, not having a clue about our precise location is more common than we may think. We have all gone through the same experience if we have arrived at a strange airport in a foreign land. Just where is JFK if you fly into New York? Or Charles de Gaulle when you are looking forward to a weekend in Paris? And when it comes to that, how many of us know where Stansted or Gatwick can be found on a map?

There was one platform I came across which made a real and unique attempt to orientate its latest arrivals. The safety induction included a short piece of film entitled 'You are here' which gave some idea of where the platform stood in relation to the UK and other platforms. At the very least this meant that a call home would give those on the beach some clue as to his (or her) whereabouts. It was dreadfully easy for this disappearing act to become even more complete if there were a last-minute switch to yet another platform! – something that could happen quite often when work was rescheduled. A man could simply go off into the blue, whilst his wife thought she knew, roughly, where he was. Meantime, her husband was busily working away somewhere else, having forgotten to let her know the change of plan. Mea culpa! I did this several times. In the flurry of a revision of plans, I found it very easy to forget to tell 'herself'. That was why so many wives had a concern when the media broke the news that there was something amiss offshore.

The press and radio kept a watchful ear on coastguard radio frequencies and it was quite possible for an oil company to get an inkling that something had gone wrong from the BBC, or a curious newspaper reporter checking out information, only to discover that he was the harbinger of bad news. The words 'reports are coming in from the North Sea' could presage something ominous and be heard with dread in many a home. It had the same ominous ring as a news item that began 'The Admiralty regrets to announce . . .'.

Maps which detail the location of all platforms in the North Sea are printed from time to time, but their circulation never seems very effective, and I doubt if they find their way into many homes. When I was doing my homework before I first went offshore, I was surprised to discover how difficult it was to get hold of a map detailing the location of all the installations. If things go wrong, newspapers are desperate for dramatic pictures, but usually have to be content with rather elderly 'stills' of platforms. Usually there is very little to indicate the platform's location. The vagueness can result in a double whammy because not only is the location unknown, but often the sort of work being carried out there is not remotely understood. In most walks of life there can be some comprehension of the way people earn their daily bread, but when it comes to the North Sea, the confusion can be quite spectacular.

There was a family living in the north east of England, in Darlington, and every fortnight the wife took her small son to the station to meet the man of the house off the train from Aberdeen. Two weeks later the procedure would be reversed and this time there would be fond farewells. This routine went on for many months and then one day the little boy asked, 'Mum, what does Dad do on a train for a whole fortnight?' Question: just why is the oil industry so ineffective in its communication with families and, it must be said, the general public? It may be good at company reports and bottom lines. But when it comes to people . . . ?

An operating company was kind enough to give me a package of their well designed books of 'cut outs', the intention being that children could make up their own model platform. I asked around only to discover that not a single school I contacted had every heard of such material. Such a pity, for so much ignorance could have been dispelled, the remarkable feats of engineering appreciated, and the lives of a workforce better understood.

There were times when I thought that those in the offshore world lived so close to the operation that they had forgotten how remarkable the whole enterprise could be to other people. There is almost a coyness about an industry which after so many decades has still failed to persuade people that it is platforms, and not rigs, which produce the oil. And, if so inclined, where can you go to educate yourself? With all the human endeavour, skill and courage as constants in the North Sea, it is strange that the industry can be content with a few miniscule rooms in Aberdeen, rarely visited, which detail the offshore industry. What a contrast to Norway, with its sensational museum at Stavanger, where great feats are worthily recorded. Why the dumbing down of something of which we should be proud?

Probably we are all rather vague about the occupations of others in our families, but at least we know where they spend their days. But offshore workers are so widely scattered that they lack the mutual support always available when there is a discernible community of fellow workers equally sensitive to the unique anxieties that can arise in an emergency. Within the Aberdeen area there is at least an awareness of the continuing activity. The routine helicopter flights overhead provide a constant reminder. But what about the wives who live in Devon, the Isle of Wight, Swansea, Stockton or Stornoway? Their stress can only be guessed at, for in such locations the distinctive offshore ethos is quite unknown.

Oil companies seemed to vary in the importance they attached to the family unit and their need to support it. There was a time when visits offshore were arranged for wives and partners. The trips were much appreciated, not least because they made it easier to understand the working conditions. Obviously, such projects did not come cheap; they presented safety problems, and they were restricted to wives of the operating companies' employees. The commissioning of a video depicting the daily routines throughout the platform was a more practicable alternative, and did its best to anticipate the questions the offshore worker's family were most likely to ask. It only scratched the surface but answered the questions that sometimes people did not like to ask because they could seem so concerned with small details. What were the menus? How was the food prepared? Where was the laundry done? What books were in the library? What videos were available? What were the cabins like? And the toilet facilities? It was the answering of such questions that gave the video its value. Significantly, very little footage was devoted to the technical aspects – it was the living offshore that aroused the curiosity!

Another attempt was made at family familiarisation; I joined a chartered train taking interested families from Aberdeen to Newcastle to inspect a platform that was being fitted out on the Tyne. It was a fun trip but for me the most valuable part of the whole exercise, was the journey itself. It gave families a wonderful opportunity to relate to each other. It confirmed my impression that there have not been nearly enough opportunities for the offshore family to get together.

So often, after an incident with tragic consequences, I found that a priority with the bereaved was to have an opportunity to meet with a loved one's colleagues. It would have been strange if it had been otherwise, for these were the people with whom those at home shared a husband, a partner, a son or a daughter. Invariably, when I introduced a widow to some of her husband's mates there was an immediate recognition – 'My husband spoke a great deal about you' – and the bonding and the healing were palpable.

It is self-evident that there is every reason to bring people, in these sort of circumstances, together. But who is it who could, or should, organise such gatherings? Is it not something that fits naturally into a company remit? Maybe it is the hole in the heart of the offshore oil industry, for although North Sea extraction is highly technical, it is also intensely human. Yet there is no group set apart to think creatively about human issues. If such a group were in existence then there would be many practical spin-offs, ranging from the frequently forgotten consideration of the human implications following a critical incident, to the fostering of the offshore family, the appropriate recognition of some of those within it, and the dissemination of the information which can interest the general public.

In the Oil Chapel in St Nicholas Church in Aberdeen there is a beautiful pewter bowl which bears witness not only to the generosity of The Scottish Wives Offshore Group, but, equally, to the spontaneous grouping of individuals with common concerns. The influence of wives and families cannot be overestimated and can be an immensely valuable reserve when things go wrong – but it has to be organised and sustained.

The Christmas Service for the oil industry is now well established and has become increasingly more imaginative (since I retired!), containing interviews with families living all over the UK. It's a significant gesture but cannot hope to be more than that in developing the concept of the offshore family. There have been times when it has been obvious that the scattered nature of the workforce has been to the advantage of those who run the oil industry. One is forcibly reminded of the old maxim about dividing and ruling. If, through circumstance, the divisions are already there, then the set-up is ripe for the control that can help to keep shareholders reasonably happy, politicians relatively quiet, and the environmentalist lobby off the boil – but where does all that leave the workers?

The matriarchal influence in industrial history is well charted. When great disputes arose in the heavy industries they could never have been sustained if it had not been for the support the menfolk received in their homes and within their tightly knit communities. The isolation of the North Sea production platforms mirrors the separateness of the families on the beach. Family units can disintegrate and, sadly, those who work offshore have not been immune to the misery of broken marriages. An explanation might be sought in the time spent in a job which claims six months in every year. But maybe the reason is rather more subtle. There are many in the Services, industry, commerce, the Merchant Navy, to cite a few examples, where the head of the family can be absent at any one time for much longer periods. Maybe it is not the frequency of the separations or their duration; but their sheer predictability that is the killer.

One day a control room operator showed me a five-year calendar on which he had marked the Christmases he would be offshore; the number of birthdays and family anniversaries he would miss, and the number of school prize-givings and concerts he would not be able to attend. The relentless predictability of this sort of life style demands forbearance and understanding of a particularly high degree – from both partners.

We are perpetually reminded that a good marriage depends on good communication, and most of us are not as good at this as we should be. Whilst marital differences can provide some of the adrenaline for a good union, it is not a good idea to have a row just before a trip begins; a fortnight apart can magnify a minor difference and make for a frosty return. In the earliest days offshore the radio telephone was the sole means of calling home. By no stretch of imagination was it confidential and the North Sea (one of the world's great gossip shops) could be agog with 'a situation' that was none of their business. Now communication by satellite means that people are no longer so incommunicado and there are the opportunities to patch things up. But it's the old story: the desire for reconciliation has to be there and sadly sometimes it is absent.

What has come to be known as 'the intermittent husband syndrome' presented special problems for those with children. This was an area that vexed many a family man because it raised the question of a child's perception of a father. Was he to be the disciplinarian who once a fortnight would ride in like some Hollywood cowboy and sort out the misdemeanours of the previous two weeks, or was discipline best left to the ever-present mother to administer as situations arose?

Another source of potential contention was the return home at the end of a trip. It could mean a more relaxed routine because one half of a partnership would be in some sort of holiday mode: eager to entertain and be entertained. Not so for the spouse whose life continued in the relentless daily routine. Keeping an even keel in this area of delicate relationships could be something of an art form. Happily, many couples were good at it.

Some men felt uncomfortable when they were unable to share fully in the family crises which could suddenly flare up – not least with adolescent offspring. The obvious solution when the situation became hard to bear was to change jobs and return to work on the beach. But it was not quite so simple because another condition then appeared which imported its own particular difficulties.

'Golden handcuffs' are not easily removed because, when they are slipped on, they are the indication that the worker has come to accept the offshore allowance as the transforming ingredient in his life style. This is reflected through the increased income that makes so much more possible. There are not many people who will voluntarily forgo life's luxuries honestly achieved. To take such a step requires the wholehearted support of everyone who would be affected. Few felt able to contemplate such a major change and there could be a lot of unresolved dilemmas both on and offshore. It's not easy to say farewell to the exotic vacations, let alone the new car.

Moving around a platform made me realise that there were a group of workers for whom the offshore family was their only family. As a result of a domestic break-up, time on shore could present all sorts of problems, not least when they found themselves totally alone. When the time came to return to the beach, they would say cheerio to their mates who were rushing for the transport which would take them to their own firesides and then there would be the sad sight of a lone figure making his way to the airport bar to sit and try to work out what he would do with the next fortnight. The companionship and camaraderie offshore meant everything. It was there that they found the unconditional friendship for which they hungered. Often I was struck by an attitude that traditionally is found in the

French Foreign Legion. People pointedly curbed their curiosity about their colleagues and personal information was rarely asked – it had to be volunteered. Some would dread the moment when they boarded the chopper for the beach where they would miss the friendship of their workmates and the sober security of the messroom. They would find consolation in the bottle. To sit talking in the recreation room with a man who hoped for oblivion soon after he reached the shore helped me to glimpse the hopelessness which enveloped some workers. But some help was at hand.

I was blessed (the only possible word) by the friendship and support of the late Sister Mary McDonald. She was a nun. Her wise eyes twinkled through owl-like spectacles and her rosy cheeks brought a refreshing air of innocence to a sometimes tawdry world. Appearances could be deceptive, because behind the seeming unworldliness there was a sharp and realistic mind that, no matter how disconcerting the experience, could look with enormous love and compassion on her fellow men and women. Her dedication was total; there was never a thought of personal glory, and her concept of service was breathtaking. Equally helpful was her wicked sense of humour, and when I would become frustrated with my inability to get things moving, she would restore my sense of proportion and make me laugh at myself. It was a friendship that was very precious to me – perhaps because in her grace, stability and religious certainty, I recognised all those qualities that I coveted, but which only God, in his wisdom, bestows. I envied Mary her radiance which she retained until, after a brief illness she quietly passed away – unassuming to the end. I shall always miss her, for she epitomised to me the 'Costly Grace' so beloved by Dietrich Bonhoeffer, the German Protestant theologian and martyr.

Mary ran a hostel in Aberdeen that was small, welcoming and comfortable. It was like a family and there were men whose lives were turned round by her unconditional care and concern. Her bubbly personality could transform many a potentially difficult situation, and her untimely death, as unobtrusive and dignified as her life, was such a personal loss. She worked discreetly and wisely; always in the background and was clear in her aims and objectives and achieved them. All she undertook and accomplished has been scarcely recognised – except by those who were most directly involved and who will acknowledge a personal indebtedness for the rest of their lives. In the midst of a lot of frenetic activity she provided the pool of calm and stillness that so many sought and, through her, found. Her contribution to the wellbeing of the offshore workforce was incalculable.

Mary was typical of the self-effacing group of men and women whose concern for those who worked offshore has placed the oil industry in a great debt. The delightfully diffident James Stewart, who ministered at St Nicholas Church in the centre of Aberdeen for many years, is but another example. Behind an apparent chronic absentmindedness, there lurked a remarkably steely will. The chapel created to honour the memory of those offshore owes its origins to the determination of James, who beavered away making certain that the best designers and craftsmen were employed to establish a very special spiritual base. And so the cast list goes on. There was Mike Marshall, a professional cameraman and film producer, whose experience of offshore life was considerable as he moved around recording events for training purposes, and maybe more significantly, ensuring there would be archive material of huge interest. From my particular point of view, he was always willing to go many extra miles beyond his contractual obligations, and on his own initiative produced a moving fifteen-minute epilogue that was distributed to all platforms after the Piper Alpha disaster. It was profound and pertinent and there were men who found this particular presentation of the Christian hope to be real to them. It was not unusual for the video to be borrowed from the wireless room and in quietness it would be played as a support in times of personal sorrow.

Of course, there were many men and women in the oil industry who gave their time and abundant talents to support the Chaplaincy. Some did so out of total conviction that it filled a gap in the resources available to the workforce. Others were not so convinced, yet instinctively felt it was a 'good thing' and should be encouraged. And there were those whose personal life style left little room for spiritual things but who were big enough to make sure that every facility possible was offered to those who thought otherwise.

Committees and Trusts were manned by very busy people who cheerfully accepted their companies' nomination and faithfully attended meetings and implemented their decisions.

Although the Chaplaincy had no financial resources of its own, everything from office premises, furnishings and equipment, publicity materials, invitations and programmes, were produced or made available, accompanied by huge goodwill. Maybe the first spur had been won?

Of course, there were individuals and a few, very few, companies, who did their best to side-line the Chaplaincy. This was unfortunate, yet a healthy sign of a country where freedom of opinion was enjoyed. There was the view that oil and religion mixed as badly as oil and water and this attitude invariably belonged to groups who were astonishingly oblivious to the human factors in the industry. Their attitudes, revealingly, were reflected in the noticeable lack of stability and wellbeing of their organisations. The offshore world was a microcosm of the world at large.

11

Power and Politics

LOG EXCERPTS

📝 *Was flown from Statfjord to Brent Charlie (4 mins). It is not hard to define the difference between the UK and Norwegian offshore ethos:*

> a) *In the Norwegian Sector people come before all else.*
>
> b) *In Norway a very long-term view is taken about the extraction of oil.*
>
> c) *Less of the macho concept.*
>
> d) *Less affected by the US myths about oil extraction that pervade the UK sector.*
>
> e) *Norwegian workforce is almost exclusively nautical.*
>
> f) *Less obvious pressure on extractors.*

📝 *From a management point of view the situation is delicate for at no stage must the labour troubles be allowed to develop an emotional content. There is no room offshore for a frustrated worker to do something 'silly'.*

📝 *Had lunch with the two men who had managed to escape from a gas-filled leg – they were still in a considerable state of shock. They asked me to sit with them as they were questioned by the Department of Energy. Found most of those to be asking the questions to be very sensitive, with one exception whose authority had somewhat corrupted him – the unattractive face of the Civil Service.*

'With our next promotion we shall be getting a knighthood.'
A devoted personal secretary (Civil Service)

'It is a worker that has been engaged, but half a life that has been appropriated.'
Oil company CEO

🗂 *If only some of the heads of oil companies – not all – could understand that often the production results can flow, quite literally, if there is a better empathy with the workforce.*

Every February the oil industry forgets its diaspora and comes to town to enjoy a week of activities organised by the Energy Institute (formerly the Institute of Petroleum). The industry's movers and shakers appear to be all too willing to respond to the annual rallying cry to make their way to London. It is a notable time, not only for the very impressive places where the various functions are held, but also for the prominent figures from the oil and political worlds who are happy to mingle in a display of spontaneous goodwill. The days are a testimony to the mutual sense of power shared by these two groups when they get together. The internationalism is self-evident, as are the remarkable relationships that are established worldwide with political structures. It is to be expected, for many multinational oil companies enjoy a fiscal and economic power that can be awesome. There is no other industry which, quite literally, holds the stability of nations in its huge hands.

The complete dependency of modern society on assured oil supplies was dramatically illustrated when a nation began to grind to a halt. UK-wide protests over the level of fuel duty resulted in an acute oil shortage and the domino effect even surprised the most militant protestors.

This probably explains why most governments keep close to their oil captains. Apart from anything else they are seen as able men accustomed to making decisions. It cannot be coincidental that much use of them is made in government commissions and quangos. This residual power has placed the British offshore oil industry in a very special relationship within its own country and on the international scene. It is interesting to contrast the attitude of the government of the day towards the railways and their operation. Privately owned train companies can be subjected to a considerable degree of control, and politicians and the media are not loath to level a caustic critique at their perceived operational failings. The attitude towards the oil industry is much more kid-gloved. Perhaps that is to be expected when the industry concerned is international, worldly and extremely influential. It exudes a confidence that comes with the knowledge that the world is oil dependent, and its wealth, power and worldwide involvement makes other endeavours seem rather provincial.

The annual Energy Institute Dinner is one of 'The Week's' events. It is a monstrous affair held in the banqueting room of London's Grosvenor

House and attended by over a thousand uniformly attired men. Fortunately the sartorial blandness is relieved by the dinner dresses of a sprinkling of ladies. After I was appointed chaplain, one of the perks was to be invited each year as a guest at a company table. In 1991 I was asked to deliver the second speech – an interesting challenge. The first utterance is traditionally delivered by a national figure of gravitas. My verbal offering was supposed to be light-hearted and intended to entertain.

I was a bad choice because I am not one of those gifted clerics with a stream of funny stories, and anyway the mood was more sombre than usual. The Gulf War (the first one) was in the offing and everyone was pre-occupied with events in the Middle East. The concern was tangible and inevitable arising from the international awareness of experienced oil men. The Foreign Secretary (Douglas Hurd) spoke of the concerns national and global and their economic consequences, with every word emphasising the geopolitical implications of oil activity and the need for military intervention to preserve the status quo. When it came to my turn, I tried to bring a much preoccupied audience nearer home by talking about Britain's forgotten army – the offshore workforce. This was a point that had to be made in an evening when everyone's thoughts were with a British Army at war in the desert, and the oil refineries in Kuwait were, quite literally, going up in flames. People at the 'sharp end' can so easily be forgotten by those who are used to involvement in the big picture.

There was one advantage in sitting at the top table: it gave a perspective on the great and the good on parade, and the favour Her Majesty and her Government accord the oil industry. Awards graced the necks of many of the industry's leaders and it was all very impressive, but I found my thoughts returning to the North Sea and the men and women on the platforms, for the occasions when any recognition had come their way were all but non-existent.

In my time as chaplain, within the Department of Energy there was a section devoted to the oil industry. It was headed by its own Minister of State with the support of a cohort of Civil Servants. The late Peter Morrison was my first sustained encounter with a Minister. He was the quintessential patrician Tory, with a somewhat disconcerting predilection for fluffy toys, which crowded the top of his desk. As I came to know him, I tried to fathom the political mind where the creation of working relationships between oil men and the political process were all important. The good minister has to be shrewd in his judgements of people – he was – and he was not loath to reveal them. At one across-the-desk meeting, I was expressing myself, somewhat forcibly, about my perception of some of the prevailing practices within the offshore oil industry. When I paused for breath he quietly said, 'Andrew, there is one certainty in life – you will never be appointed to a quango.' He was good at determining who could best work together and then making sure that they did get on. Wheeling and dealing can be an art form, as is the ability to make a stand and yet leave a door open for an emergency exit.

For me it was a very new world and one that required an agility of mind which, from a distance, I came to respect. But membership of a government means being tarred with a particular brush and a change in leadership has an effect that cascades down the chain. When John Major succeeded Margaret Thatcher, there was the inevitable ministerial reshuffle. Peter Morrison was a big, florid man and he was followed by his physical antithesis. Colin Moynihan had the stature befitting a boat race cox and it was not unknown at an oil dinner for some well lubricated guest to call out once the new Minister started on his speech, 'Will the speaker please stand up.' Moynihan was a gifted individual: a boxing Blue and most talented pianist. Within his compact frame was compressed a prodigious energy. I found it fascinating to watch the apparent ease with which he mastered his new and complex brief. The Ministries of Sport and Energy were both about the explosion of power but in very different spheres, and one saw how opting for public life was no sinecure, producing the uncomfortable paradox of being at the centre of attention and yet living with a considerable loneliness.

Political and industrial leaders had their own particular gifts which I came to identify. My curiosity was aroused and bit by bit I found myself trying to pick out the talents which made the leaders in the oil industry able to bear their burdens with such equanimity. The answer did not come easily for they were a surprisingly disparate group – just as with ministers of religion there was no obvious 'type'. Take these three very contrasting individuals. The cigarette holder of John Brown went well with a dapper, immaculately clad individual. A Trusteeship of the British Museum hinted at interests that were far removed from North Sea platforms. Bob Reid made a point of shoving his office desk against a wall so that all meetings and interviews were held seated around a coffee table. The fact that he had lost his right arm when a lad working in his father's butcher's shop never appeared to inconvenience him or anyone else. Sam Laidlaw was a devoted family man, and all his travelling and his enthusiasm for ocean racing (for a time he was posted 'missing' in the notorious Fastnet Race) concealed a sensitive and reflective nature.

These three men came from widely different backgrounds, were all exceptionally able and with a professional competence you took as read.

They were oil men throughout their working lives, yet there are many other captains of industry whose personal qualities are not dissimilar. So what made them so special? I found it interesting to look at their eyes, for they had seen more of the planet earth than most people and this gave an extra dimension to their thinking. They wore responsibility as a natural garment. In contrast to so many others, they were far beyond the stage of being personally ambitious and relationships with them were free from the calculation you detected in lesser mortals. I enjoyed my contacts but regretted limitations of time prevented the fullest development of discussion with them on the key issues of the oil industry which, it seemed to me, related to the environment and social responsibility.

It was probably a utopian dream, but I always hoped that at least one of the great companies would develop an imaginative 'human responsibility' department where the implications of being producers of oil could be considered in detail. Shareholders had not only to be kept happy – they also had to be educated.

With great companies should come the recognition that there are areas of moral obligation that go along with the power and prestige. Excluding the self-evident need to satisfy shareholders, in the unique circumstances of the offshore oil industry I would place a responsibility for the workforce as the prime priority in its internal operations. Of course, good food and good wages are important but there is far more when expecting people to give their lives to an extraordinary style of existence; knowing separation from loved ones on a relentlessly predictable basis demands more than fiscal reward and quality nourishment. The moral obligation can only be fulfilled when the best communication is established between family and company, and the second home – the platform – is designed and equipped in a way that shows thoughtfulness and sensitivity to prevailing demands. Macho traditions die hard, but a new era in offshore activity has dawned. There are greater expectations from a changing workforce which will be working in increasingly harsh conditions and more, well qualified women will be competing with men for appointment. The implications of this constant pressure on the glass ceiling are significant and a worthy response demands some reflection. Only the leaders have the power to initiate requirements unique to this industry. It shares a moral responsibility for health and safety and environmental control along with many industries but it is in its residential responsibilities to its workforce that it is unique.

Two very different organisations dominate the oil industry in the UK. The Energy Institute is apolitical. In some ways, with its local branches it is reminiscent of a network of chambers of commerce, with the difference that it seeks to bring together the manifold concerns of one vast industry. It is saved from becoming a dining club through the formulation of imaginative courses and lectures designed to meet the technical needs of an industry both up and down stream. And it presents a global view of the industry's activity and draws attention to the latest legislation and advances in research and development. It is not so much an initiator of new ideas as an encourager and developer, and in a very real sense is the bridge between the extremes of the drilling platform and the petrol pump. Inevitably, I find it disappointing that little space, if any, is devoted to human factors. After my final offshore trip, the Institute kindly invited me to write a piece for their monthly magazine. My effort was duly printed but there was no reaction to it. It would have been good if someone had written to say that there was no place for a 'people' article in a technical magazine, or that I had pointed up something that was all too easily forgotten. But it was hardly surprising, for soft issues are given very little place in an industry that prides itself on a 'gutsy' approach to its operations.

The United Kingdom Offshore Operators Association (UKOOA) is very different. Discreet in operation, it is the power-house of the upstream side of the oil industry. It is no chamber of commerce but came into being to bring greater understanding between organisations who were all engaged in the same activity. I found some of the parallels with international church bodies to be uncanny, with the rivalry, the secrecy, the wielding of power, the conditional unity, common to both. To have, at least, some things in common can be a good thing and makes sense of some aspects of offshore operations. One could not fail to notice that although the operators worked in a highly competitive and confidential milieu, they could produce a common front when government legislation was being negotiated and taxation being determined.

In a most practical way UKOOA made the Chaplaincy to the offshore oil industry happen. It can seem churlish to appear to bite the hand that feeds you, but for all its support paradoxically it, too, is weak on 'people' issues. I used to feel they were happy to encourage me to develop my own thing, even if, for many, they were not too clear what that might be. When I was particularly tired, I sometimes felt that there had been a transfer of responsibility, never discussed but tacitly assumed, that if the Chaplaincy did some sort of job then the difficult area of people was being dealt with.

Keeping a good relationship with one's paymaster could be a delicate business, and on the whole it stayed on an even keel, but all the time I sensed that the people issues which concerned me were a bit of a mine-field that had to be handled with care. In the early days I had a memorable meeting with a UKOOA committee, of some size, who had taken issue with my published description of the practice of what I described as 'industrial apartheid' offshore, where the relationship between company and contract personnel had resulted in the emergence of the self-styled 'tartan coolies'. Matters were sorted out but there were times when it was a bit like treading on eggshells. I was always mindful of the privilege in being allowed to go offshore, and one that could be easily withdrawn if I was deemed to be a tur-bulent cleric. It was a learning process, for both parties, and through time I discovered how I could avoid compromising myself and at the same time try to ventilate issues which mattered. Maybe any worthwhile chaplaincy is balanced on a knife-edge. The art is in retaining one's equilibrium.

The politics of dealing with the workforce were as complex as the work-force itself. In times of unrest there were three distinct groups involved. Operating company employees worked within the normal parameters of industry with a career structure, pension rights, medical oversight and employment protection. The contract worker could be seconded to one par-ticular company for many years, yet remain all this time outwith the support systems administered by the operating company for its own employees. And there were the itinerant workers who went offshore with their special-ised skills to carry out specific tasks and then move on. The operation of a three-tiered workforce was further complicated by its physical location and its dispersion when ashore. Some disputes were mischievous – the sort that arise when monotony takes a grip – other issues deserved to be heard, but there was no voice which commanded attention and no ear to listen and react with credibility.

I always had to keep on reminding myself that working offshore was an exercise in choice, and that there were those who had deliberately opted for that way of life because it freed them from the union activities they had come to detest in their previous occupation on the beach. But when there was unrest offshore, it was very necessary to have machinery for airing grievances, for it was not a location with room for unrest.

When I worked in Inverclyde, the union tradition was bedded into work life but I always remembered an outstanding union leader remarking to me that companies got the unions they deserved, and it was noticeable that there were companies where union representation was far from tiresome and did much to help management to achieve a smooth, well-ordered, har-monious operation.

I was never able to fathom the deep suspicion of unions held by many of the oil companies. I suppose what is unknown is always more threatening, and for all the outstanding ability in the company management, I was struck by their relative inexperience in labour relations. The management of oil wells and refineries in distant corners of the globe is not the best preparation for the rigours of labour relations in the Western world. Yet the areas from which most of the North Sea workforce was drawn were those very places which spawned trades unionism and traditionally where workers were used to having a voice. Grievances can occur, misunderstandings arise, and the channels for their ventilation can so often pre-empt the upsets that are damaging for everyone concerned. And yet . . . the offshore scene is unique with myriad levels of work relationships; a twenty-four hour day involvement; a workforce that is widely scattered for six months of the year and which rarely, if ever, meets up with those who do their job for the other half of the year. Add to all that the fact that employers, many from other countries, were not overfamiliar with the work ethos of industrial Britain, and you get a very unco-ordinated body. The one unifying factor is the shared experience. At the end of the day, it seemed to me that in practical terms an umbrella union to represent these disparate ingredients would not work. But everyone was entitled to a voice and the only workable channel could be through an ombudsman thoroughly familiar with offshore technology and the working conditions of those who made it all happen.

The ability to communicate, the awareness that made certain there was always time to show interest and concern which could transform a culture from blame to praise, the capacity to listen and understand a different point of view, and the readiness to like people, are all so obvious when listed. Yet these were the qualities that could transform a work place offshore as much as on the beach. However, on a platform there was an extra dimension to management. The workforce lived on the site as a compact community. No-one could down tools and leave on the spur of the moment and it produced an extra raft of responsibilities in terms of welfare, wellbeing and safety.

I am no ombudsman, but in my own quaint way did my best to represent the interests of those who worked offshore. I had none of the practical skills, but that may have been an advantage because I was able to interpret situations as I saw them, without being burdened with all the technical baggage which can be a mixed blessing. On one particularly savage day, I was crossing the top deck of one of the more modern, flatter platforms, comfortably secure against the elements in my company-issue over-jacket that was lined, weather, wind and fireproof (it still is wonderful for gardening after all these years!). A figure suddenly appeared, looking like a refugee from a Russian gulag. He had been working outside and was obviously extremely cold, and I was shocked to discover that protective clothing was only issued to company employees; contract personnel had to fend for themselves. Unfair? Unjust? Yet when the matter was raised, in that classic phrase, there was 'no problem'. It was as though no-one had thought about it, and in my opinion was another instance of the reality that when busy powerful people are engaged in major concerns, the soft issues, with all their irritating demands for attention to detail, are the ones that go to the wall. Another example: I spent a day of mind-numbing boredom at Sumburgh airport when the North Sea haar was at its thickest with all flying suspended. Dozens of men had nowhere to go, and nothing to do. Within a few weeks the upstairs cafeteria was provided with a large screen television projecting the latest offerings of ITV and BBC. Just two samples of what must seem to be micro detail in the total scheme of things and yet disproportionately crucial in terms of morale and wellbeing, and serving as a safety net when irritated individuals can seek to cause unrest, or be so preoccupied that they are careless in their attention to safety procedures.

Maybe there is a correlation between power and safety? The oil industry is so vast and the global implications of all it does are so huge that there is a tendency to always 'think big'. Whether it be a billion pound development programme in the North Sea or the construction of a three thousand kilometre pipeline linking the wilds of British Columbia to Chicago, the logistics can be so mind-boggling that it is all rather overwhelming and that non-technical, unspectacular, and (in oil terms), physically insignificant ingredient called man can be pushed to the sidelines.

It is fascinating to look through the large, usually red, files that deal with 'incident response'. Hundreds of pages are devoted to dealing with every imaginable engineering hiccup which fertile minds can dream up and they are followed by a few pages (in one case it was half of one page) devoted to the human implications of an incident and the best ways of coping. The decision-makers, policy strategists, and power-brokers seem to be quite clear where their prime responsibility lies in the production of a resource the world cannot do without. That is beyond dispute, but there is a case for those who extract this resource to figure somewhat more prominently in the scheme of things. My links with the Royal Navy for a few years provided a contrast in the care, support and understanding of those committed to a particular task. My experience tells me that it's the happy, enthused staff who know their welfare and interests are being respected, who are proven to be the most effective unit.

Environmental issues come to rest at the door of the power-broker, for pollution is a perpetual challenge that will not go away, and is but one of many issues that confront the decision-makers, and every response is weighed by a curious media and an even more concerned general public. There will be innumerable challenges when the decommissioning of platforms begins. A pilot run took place when the Brent Spar was removed. It created headlines and was a reminder of the ability of the great British public to adopt a moral standpoint on precarious information. Equally significantly it was a reminder that the offshore industry can be subjected to intense public scrutiny. Mass moral indignation, not necessarily based on accurate data, appears to be a reality that the industry has to be prepared to thole. And this arousing of public moral outrage covers many things, for it's not just the environment, but human rights which can be involved and the championing of the under-privileged. Indeed, wherever the industry operates it is increasingly immersed in some of the issues that arouse emotion and strong reaction. Maybe every corporation should have its own tame moral philosopher, for there are challenges within and without that cannot be avoided.

It is natural to speak about the 'power' of the oil industry yet it is hard to quantify it. Just how does one calculate the influence that comes with the possession of something on which the whole human race depends? In normal marketing terms the sale of the same product would result in intense competition and a lowering of prices. The oil industry is different and unique, with its high level of competition and yet with forecourt charges which show little variation. Companies that are household names the world over, face each other across the road, clad in their company livery, displaying minimal price differences. The paradox of competitiveness, yet mutuality, is just one more indication of the way the great power within this global industry has produced a system that irons out differences and proceeds on the basis of agreed understanding. When OPEC meets it holds the whole world in its hands; power and politics intermingle. When oil people discuss issues they talk in telephone numbers, such is the size and complexity of their operations and the cost of their implementation. The comprehensive nature of oil extraction makes this planet seem very small. It all seems a far cry from the sharp end, where those who produce the black gold make all the wheeling and dealing possible. However, the one cannot do without the other.

12

The Lessons

**'You're kind of different
done up like that.'**

Offshore worker to chaplain after
a church service on the beach

LOG EXCERPTS

*An astonishing two hours (!) in the Bear Pit.
We discussed the World Cup, the Oil Chapel,
Rangers/Celtic, the Yorkshire Dales, Industrial
Mission, Offshore Safety, Offshore Unions, the
Church, and UK and US cultures in the oil
industry. Everyone seemed to have views on
everything.*

*Met the five full-time Norwegian chaplains
in Bergen. A wide-ranging discussion where
we discovered our problems were similar,
opportunities are enormous, and we shared
the same vision of the Church. We agreed we
should meet, at least, on an annual basis – we
have much to learn from one another.*

*Out of the blue someone remarked, 'If more
ministers did what you are doing, folks would
want to come to church.' Allowing for a wee
bit of optimism it still has to be true that we
must, first of all, be where people are, relate
to them and be interested in the 'where' and
'how' they earn their daily bread. So much in
the Gospel can make sense once people are
reassured that they matter.*

*Pastorally speaking an interesting trip. Met a
Dutch driller, whose father-in-law is a Church
of Scotland minister and a steward, who is a
keen elder of the Kirk. Why, oh why can't the
Church, not least in this computer age – make*

a list of the vocations and locations of its membership?! The amount of ignored resource is colossal.

✐ *My presence always prompts conversation towards spiritual things. There seems to be a lot of space in men's lives for religion – not so much for the Church as it is presently perceived. Once the offshore Chaplaincy was under way, it highlighted the remarkable credibility gap that could exist between a religious institution and the people it purported to serve. It is exasperating to hear a Church on the one hand lamenting the steady decline in its membership and on the other pleading for greater support. Seemingly, there is a failure to understand that the majority of people have little conviction about institutional religion and consequently have not the slightest sense of obligation towards it. Not unreasonably, they just do not see why they should support something that, so far as they are concerned, is not for them.*

By nature most people are very pragmatic. They can immediately understand the need for guide dogs for the blind; kidney dialysis machines for the irreversibly sick; palliative care for the terminally ill; the wretchedness of starving children, and the desperate needs of families without homes. In all these things there is a need to be met. The uncertainties arise when they consider the strange building on the corner whose doors remain closed for six days of the week, and are only opened for a few hours on the seventh. What is the need it seeks to satisfy?

From time to time someone working offshore would try going to church and then report back. They could be genuine enquirers or merely curious, but no matter what propelled them through the doors they invariably discovered that with the best will in the world they could make little of it. The common experience was that they felt they had been harangued from a pulpit (the traditional six feet above contradiction) and there had been a use of language and verbal illustration that was totally foreign. To make matters worse, this was endured whilst sitting surrounded by a decor which was usually cheerless and reflecting an era long since gone. As far as the music was concerned, it was deemed to be so old fashioned that it constituted a bit of a joke.

It was inevitable that whenever I went offshore a huge amount of time was spent trying to correct misunderstandings. More, personally, agonising was the picture I began to perceive of a church that could be strong in judgement and weak on forgiveness. Altogether it was a rather bleak image that emerged of a Loving Father! Yet it did not have me hurling myself off the helideck; I realised that, once again, I was rather slow in perceiving the true situation, yet one that was, from my perspective, not without hope.

The more time I spent with men and women offshore and shared with them, at their request, in simple acts of worship, the more it became apparent that there are so many now outwith the traditional expressions of the Christian Faith who have severed a once traditional personal allegiance. Their reason is not because they have rejected the Truth of what it is all about, but because they want a meaningful link in spirit with One greater than themselves. This quest is quite different to the superficial religious observance that can be unthinking, unexamined, formulaic and routine.

Time and again I was reminded of the dormant hope within so many that has been awakened by situation and circumstance and seeks relevant expression. I suspect that this has always been the case and that many opportunities have been lost. If the prevailing institution seeks to fulfil the divine commission and be the saviour of all, then the deployment of its professional representatives has to be more imaginative and reach out far beyond the protective arms of Mother Church.

Maybe it was because of my watery life but I found the temptation to liken the Church to a swimming pool to be irresistible. It's fine if you can swim, and if the temperature is OK, but singularly uninviting if you can't. What can make it worse is if there is no shallow end then anyone entering it can immediately find they are out of their depth. Sadly, this is the fate of so many who end up drowning in their incomprehension. All the time I found myself wondering where the sensitivity and grace towards those not dead to spiritual impulse and longing for something that 'added up' was to be found. These were the people who, rightly, resented being classified as 'secular' simply because they had no formal attachment to a traditional institution. If that overworked word 'outreach' (much used by the Church, since it dispensed with 'mission' in the ecclesiastical vocabulary) is to mean anything, it must surely imply the establishment of a spiritual relationship with people, in the situation with which they are most at ease. This can mean that the place of work becomes just as significant as the home. And nothing should be rushed, for the eventual introduction to a more church-centred belief requires infinite patience, and a gentle, understanding preparation before a meaningful entry.

So often I would return from an offshore trip and feel that an inordinate amount of time had been spent in doing a PR job for the Church. It became increasingly clear that the Chaplaincy task has as much to do with dialogue and explanation as with mission. Maybe that is the reality of mission in an age of communication and spin? For myself I never had a compulsion 'to bring people to Jesus'. My personal theology was my belief that God is within us all and the Master showed us what he was like. It was not a matter of trying to effect an introduction that could result in a meeting of hearts and minds. Rather, I saw my role as opening up those same hearts and minds so that people would discover for themselves the reality that lay within.

Uncritical assumptions can be quite remarkable and not least when it comes to religion. In the early days when I was making my first visits to some platforms, I would be provided with a guide who was recognised as being 'religious'. Unfortunately, rather than facilitating my entry into this alien world, it could produce problems. Invariably, my mentors were kindly, earnest men, yet with a personal commitment which seemed to leave little room for humour and the harmless frivolities of life. Their sincerity, which could shine like a laser, rather scared most of their colleagues, who seemed to think that their own souls were best saved by keeping their mates at arms length. It all could become very confusing for those who wanted to believe that that the garment of faith could, and should, be naturally worn. I have to confess that I was rather more at ease with those who were not in the habit of glancing at their wrist-watch and telling me, to the precise second, how long it was since they had been converted. Patently this very precise life change was just waiting to be misunderstood, and there were times when there were big demands on my tact and patience to handle such situations. It was a delicate matter keeping the balance between the understanding of God as the moral arbiter with expectations that verged on the superhuman, and the insight which revealed his understanding of, and love for, men and women in their error-strewn path through life.

Always there were eyes watching, and noting anything that might be deemed as deviant behaviour – especially for a chaplain. This came home to me one day when I was having a meal in the mess room and as I was leaving, a Bear came up to me and said he had noticed that I did not seem to have said Grace before I ate. I am not sure what these watchers would have made of one particular occasion when I was waiting in the departure lounge of a platform before moving on. A few of us were chatting when in burst a young man full of the confidence of youth. He was new to the platform and he had a video in his hand. He slipped it into the machine used to project the safety briefings, and having put his feet up on the chair in front he settled back comfortably to enjoy the show. Suddenly we realised we were being subjected to a particularly hard porn movie. It was gymnastic and predictable, and the reactions of those around me were far more interesting. They ranged from the incredulous to the embarrassed

– more I suspect on my behalf than anything else. These men made no moral judgement, but in their view something which reeked of exploitation and humiliation was not something to which their chaplain should be subjected. More importantly – why should anyone be so afflicted? We had an interesting discussion prompted by the unexpected visual aid, which came to an end when the chopper arrived and a more mundane video reminded us once again of those more fundamental drills of escaping after ditching through a window and before inflating our life-jacket. It had been an interesting incident. To press the eject button and stop the movie might have been expected of me but, rightly or wrongly, my snap judgement saw the incident as an opportunity to get a group to examine the value systems (or lack of them) which lay behind such a production.

I was learning! It became very clear that I had to pay far more attention to the thoughts and ideas of men and women who did not find it easy to express strongly held views. I began to see the disproportionate influence which could be exercised by the more articulate. More importantly, I came to recognise the finer feelings that lurked behind so many unprepossessing exteriors – life offshore did not encourage sartorial elegance, and even a fresh pair of overalls could fight a losing battle with weary, unshaven features. It would be a gross exaggeration to rhapsodise over offshore man as though within him lurked the finest feelings. Here were 'ordinary men and women doing ordinary jobs, only the location was extraordinary'. But there was a huge reservoir of kindness, and in many there lurked a genuine spiritual spark.

It soon became apparent that anything I might consider as being relevant to the wellbeing of the offshore worker was of little importance when put alongside the wishes of the workers themselves. In other words, after years of trying to cram people into my own agenda, I had to really listen and take on board what was being said. For me it was quite a culture change to make relevance a priority!

When I had been working as a parish minister, I had lamented with the best about declining numbers at public worship and had even been known to mutter about the raging secularity of modern man. A good example of ecclesiastical pique; because in all probability it had been prompted by people who were unwilling to fit into some scheme I had devised – without consultation. I realised that to talk about secular man could be more of a criticism of the person making the comment than anything else. Because men and women do not have an interest in the institutional Church does not justify the implication that they are disinterested in the things of the spirit.

There were many occasions offshore when there were vigorous discussions on life's great verities. These arose quite spontaneously, with not a syllabus in sight, and I would have given anything to have enjoyed such vigorous verbal exchanges when trying to fulfil the more conventional clerical role. It dawned on me that we have come to place so much emphasis on an institution that we have become trapped in a Catch 22 situation, where we continue to recruit and train men and women to fulfil roles which are patently outmoded, and the ecclesiastical handcuffs become clamped tight because of the demands in time, and finance that are required to keep the prevailing institutional show on the road. Time after time I found myself repeating the words of D. T. Niles, the eminent Sri Lankan theologian who was forever declaiming 'The Church ceases to be the Church unless it is in a continuous state of mission.' As a young man I had interpreted this to mean that the Church should always be reaching out from an identifiable base. It was not until I worked offshore that I truly understood the full implications of what was being said.

Of course, there must be places of worship where supportive fellowship abounds and the presence of the Holy Spirit is tangible. Those especially gifted can, through word, prayer and praise, lead others to a greater understanding of God. But there should be large clerical battalions who spend much time in listening, challenging and supporting roles where the human family seeks to work and earn its living. I do not see this to be remotely radical, merely self-evident and the only way ahead. The spiritual need is there and remains unsatisfied.

'Image' is a bit of a buzz word, but it can produce problems. The general view of the clergy is rather like a cocktail. Take a dash of Jane Austen and mix with a splash of Trollope; a hint of Rikki Fulton's Mr Jolly and a wisp of Derek Nimmo; mix and pour into a container laced with John Knox. Whatever else it might purport to be, this concoction is as unappealing as it is irrelevant: good for a laugh but little else.

The great challenge in any chaplaincy is not to overreact to the confused views held by so many. It is always a temptation to try and show you are as human as everyone else and then in an exaggerated gesture go overboard in the attempt. The hearty, oh so human chaplain can diminish himself and everyone is a loser. The filthiest and most inappropriate after-dinner speech I have ever heard was delivered at an oil function by a distinguished professional woman, who was only too conscious that she was the lone female amongst a couple of hundred men; she was determined to show she was 'one of the boys'. It has to be said that 'the boys' were a bit bemused.

This same trap is set for unwary chaplains and I stuck to the path of listening and learning.

Inevitably the chaplain was associated with situations where something had gone wrong and there had been casualties, but the bias could be restored by being around when everything was going well. There were times when your presence seemed to prompt questions. 'What's gone wrong?' 'What's the bad news?' 'Is there something I should know?' It was a pleasant change when people could be reassured that there was no hidden agenda and you were with them for the sole reason that they mattered and were important. I learned that the link between work and home was important. Wives and partners came to know about the Chaplaincy, and the Offshore Wives Group was a significant indication of some latent influence within the workforce. Its birth was a direct result of the Piper Alpha explosion and a deeply felt unity amongst the women on the beach, forged through a shared experience.

I found it something of a paradox that, for all the togetherness, a common sense of identity was less evident offshore. Is it not unreasonable to expect that a North Sea saga which has extended over three decades would have given birth to its own distinctive culture? Platforms do not lack their poets, musicians and artists, but the songs and stories are not there. It suggests a surprising lack of social cohesion, and hints at a minimum investment in the soft side by the industry. The UK Energy Minister of the day, when addressing the seventh International Energy Forum in the Saudi capital Riyadh, uttered words that could only have come from the pen of a professional speech writer: 'This forum has served as a great showcase for British skills and expertise in cutting-edge oil exploration and extraction.' The skill and expertise comes at a cost and so often I find I question how much this has really been understood.

A prevailing insensitivity is highlighted by the small amount of time spent in resolving the perennial question of how men could disengage from their offshore work with dignity. 'One day I'll reach the roundabout at the heliport, go right round it and return home,' was a sentiment expressed to me a good few times. I came to the view that there should be a compulsory retiring age for working offshore; something similar to the police and the armed forces. Everyone would then know where they stood and the problems presented by an ageing workforce would be eased. Too little time seemed to be spent considering human investment, in contrast to the intensive concentration on technical investment. On one memorable occasion, after the Chaplaincy had become firmly established, I was invited to meet with the senior management of an American oil company. It was notable because it was the only occasion I met with real hostility. The group wanted to express their corporate incredulity that there could possibly be a place for an Oil Chaplaincy. I do not think it was religion that presented the problem (some of them probably attended their local church), but their view of life was tightly compartmentalised and in their world there was no room for what they saw as 'do-gooders'. If they were to permit me to go offshore, it could be seen as interference in the personal freedom of their staff.

My time with the oil industry was a time of enlightenment with my eyes and mind opened to the extent of employer responsibility, which encompassed office and offshore staff, health and safety, environmental political and economic issues, the procedures for extraction and refinement, and the ultimate sale of a product the world cannot do without. Allied to all of this was the need for acceptable bottom lines and a satisfied group of shareholders. In this mind-boggling complexity of operation it is easy to see why the soft side of the industry could be reduced, with the bread-and-butter of human resource activity being solely concerned with pensions, terms and conditions of employment, redundancies, and staff recruitment. The morass that is offshore activity raises a crucial question. With the increase in contracting out, where does the operator's responsibility for those offshore begin and end? When everything goes well, that is an issue that does not raise its head but if there be a hiccup of some magnitude, human dilemmas arise which demand immediate answers. Can they be given?

13

Being There

LOG EXCERPTS

Joined by the platform manager and we had a comprehensive discussion about my work and what it might be achieving. The strong impression was that the visits were important in the lives of those on the platform. All indefinable yet it seemed to confirm that somehow, somewhere, there was a bottom line and yet it would never be clearly revealed. He also serves who only stands and waits.

Evening spent in very considerable discussion with one group who, surprisingly, didn't know very much about one another yet were in each other's company twenty-four hours a day, for a fortnight at a time. Much cheered by a man who said it was very important for me to keep coming out; it meant more to everybody than I could ever be likely to know.

Joke of the visit: the young woman on the platform, a geologist, after speaking to me, thought I was a company employee who, amongst other responsibilities, was the official 'ashes scatterer' when this became necessary. Though – in one way – she wasn't so far off the mark.

After this visit I am fully convinced that my task is alongside the offshore workforce. Either by being with them, or responsibly representing their best interests on the beach. When there is an 'incident' the support for next of kin can verge on the suffocating – quite different for those who continue to work over the horizon and can very easily be neglected.

The usual tour round the installation, no two are alike, and there were the predictable reactions as I moved about. It varied from a very warm welcome, to an interested curiosity, to a frank incredulity – very rarely have I met hostility.

Took the Moderator of the General Assembly of the Church of Scotland (The Rt Rev James Whyte) offshore. He was clad in disposable white dungarees which gave him a faintly angelic appearance.

Met with the OIM. An interesting man of no particular religious persuasion, but an open mind. Quote: 'Oh Jesus, what a pity you have to leave so soon, I've never had a chance to talk to a Father.'

Frequent references to the offshore allowance – (the first time this has happened) – presumably to elicit the fact that I do not get one.

The OIM is a pleasant man and we had a very frank talk for a first visit. He expressed the view that (a) My arrival on the platform could suggest there was a 'problem' (b) My departure could result in matters being brought to light that were best left in the shadows (c) Anyway, there were no problems!

I'm becoming very stale. Not inclined to talk to people, or to put them at their ease. All I seem to hear is the whingeing. Got to bed early – for the first time not really interested in anything outside the cabin. There was a knock on the door – an 'urgent' envelope had come over on the evening shuttle from the Eider. Inside was a beautifully printed certificate commemorating my first visit to the platform. I am deeply humbled.

The oil industry is result-driven, not just because it wants maximum profitability from its efforts, but because its whole ethos is testosterone-charged with macho types involved in a highly competitive world where decisions made can have not only huge financial implications, but possible major political implications, all of which have to be sensitively negotiated. It's a world chock full of paradox where, in a global activity, the white-collared can outnumber the blue; where the offices can reflect the latest in contemporary design and yet where the basic extraction techniques have scarcely changed in a hundred years. It is a world populated by those skilled in operating the latest technology and yet dependent on the expertise of those who can interpret geological formations established millions of years ago. It is very much a male world, with relatively few women successfully breaking through the glass ceiling. It makes use of the most sophisticated equipment and yet in some crucial areas is still dependent on raw human physical strength. It is an achiever's paradise, and the world cannot do without its product. But it is a moody industry. At times it can behave like a manic depressive, but it seems to accept its volatility and be able to accommodate its roller-coaster life style. Whether such a pragmatic industry with its preoccupation for task orientation, and precise job definition, would find it easy to accept someone with a very different work profile was problematic. When the Chaplaincy began I was openminded about its future and curious to discover if the life's experience I had tucked away, together with an intense personal conviction, could be applied to a very singular activity.

When I was a chaplain in the Naval Reserves my activities had been rooted and grounded on the statement in Queen's Regulations and Admiralty Instructions that 'The Chaplain shall be the friend of all on board.' The task that lay ahead involved rather more than being 'nice' to everyone. As far as I was concerned, its mainspring had to be the belief that every individual mattered to God and that in his sight we were all equal. For me, and these dynamics are extremely personal, I could only make sense of God's love to his human family if I was directly alongside people in their living and being. It was necessary to be where people were.

Most of those who worked in the oil offices on the beach were very different to their colleagues offshore and, sometimes, could show an astonishing lack of curiosity about life over the horizon. But amongst them was a cohort of oil men with many years of experience in the industry in every corner of the globe. This veteran group had a genuine, and very understandable, difficulty in reconciling a Church presence with the world of work they had known all their lives. But they had the flexibility to keep an open mind and see if the new arrival could justify his presence in this alien environment. It was no mere coincidence that many doubts were dispelled when a Chinook helicopter crashed into the sea off the Shetland Isles and many lost their lives. Also at this early stage, I had to make sure that the Chaplaincy could be understood to be more than an assistant to the undertaker.

The complex human issues which can emerge at such a sad time were starkly revealed, as were the deficiencies in coping with them. Quite suddenly, the difficult to define, but crucial role of listener, empathiser and articulator of feelings, as well as counsellor on matters to do with life, death and bereavement – together with the injection of a good dose of Christian hope, became more clearly understood. It was a significant step forward and it paved the way for a warm, if at times uncomprehending, acceptance. Chaplains have to be seen as relevant in the good times as well as the bad, although when everything is going smoothly there can be problems because when everyone feels secure, and self-sufficient, the need for the chaplain is less evident. For me there was always the challenge of showing that God is for Monday and not someone who is conveniently boxed into a religious slot in a designated building on one day of the week.

As a parish minister it had not taken me very long to realise that behind every set of curtains in every street, some sort of drama was unfolding. Not necessarily headline grabbing, but those all too familiar domestic situations which spawn a whole range of feelings, ranging from agonising unhappiness, or deep personal anxiety, to a huge sense of fulfilment. There is no escape from life's rich tapestry – it can enfold us at any moment. There can be sombre threads of pain and loss; golden ones of love and laughter, and it is all set against a background that is an amalgam of light and shade. As I did my rounds of the platforms, I listened and I grew in understanding.

It may not seem to be important to know about the quality of the bowling greens in Inverurie, or the challenge presented by deep sea fishing off Brighton, let alone being able to distinguish between the respective merits of Butlins' at Ayr and Minehead. But, what can be trivia to one person can mean a great deal to someone else. By trying to understand another's concerns, I discovered a reciprocal interest in my own and there was much home-spun wisdom in the mess room that stood me in good stead. It reminded me of my days in Switzerland, where the majority of my fellow expatriates came from the States, and I would scour the pages of the *Wall Street Journal* to keep up with the developments that could affect their working lives, and in return they would offer sage comments on faith in a competitive society.

As everywhere else, those offshore could quickly detect insincerity; it took time to gain trust and understanding – there were no short cuts. Living with the crew meant that one was subjected to a prolonged scrutiny and I had to rapidly get rid of an inflexible professional mindset. It was not a matter of a visit to a platform being my opportunity to get to know everyone. The boot was on the other foot and it was the chance for a crew to get to know me. They did. It was very precisely summed up in a skilfully drawn cartoon given to me on a platform one Christmas Day. It was shrewd and salutary in its summing up of the chaplain. He was depicted as a rather portly figure lying at ease on a cloud, smoking a cigar with a glass of wine in his hand. A large seagull is about to land on the cloud holding an autograph book in its beak and saying, 'It is you, isn't it?' Accurate perceptions are humbling and very good for the soul. I cherished many frank comments that were made about me and what I was trying to do. They were not malicious, and were a constant reminder that I was as frail and vulnerable as everyone else.

Just what was I trying to do? To be a friend, when necessary to be a confidante, and a strength and stay in times of trouble and all the time there was that extra ingredient where the reality of the unseen world of the spirit might be evoked and examined. It could be a lonely business and surprisingly exhausting. It was not just the draining process of listening acutely to individuals, but the hours that were involved, where day and night shifts had to be contacted and movements between platforms accommodated at most unsocial hours. I found it hard going, and there were times when memories of the years on the lower deck, and the stultifying monotony of a monosyllabic vocabulary came back all too vividly. There were times when I wearied at the coarseness and brutality of expression, and then I would be ashamed of myself for being so uptight with those who had not had the educational opportunities that had come my way – through accident of circumstance, and not merit. And all the time there could be a vague sense of entrapment induced by a steel monster that never went anywhere and that you could leave only when your job was done.

In the early days, seemingly insignificant gestures meant a lot to me. One supper time I was queuing for my food when there was a great yell, 'Andy, over here,' to discover a team of scaffolders had kept a place for me. Daft? Maybe, but I felt I was being accepted as part of the team, and through time this sort of experience increased as men and women felt able to share their friendship, as well as their hopes and fears, with me.

As my understanding grew, I would try to put myself in the position of the man who lived in dread of his offshore trip coming to an end and the return to the beach each day getting nearer. Happily, such situations were rare, but there were a few good men and women who found offshore to be something of a refuge from the turmoil of their personal lives.

Those whose marriages were under strain could be presented with an acute personal dilemma. It was not unknown for the regular, even, flow of the work routine to make the offshore life surprisingly attractive. Communication from the beach was not necessarily easy, and the resulting isolation could prove to be a blessed relief from the bickering and hassle which so often can be the indication that people are unhappy with one another. But absenteeism from the domestic round had a downside. There could be an unwillingness to face up to difficulties, and absence, and lack of communication, compounded misunderstandings.

Crews could remain together for many years, and whilst everyone came to know each other very well, as individuals, they did not necessarily know much about their colleagues' life on the beach. Idiosyncrasies were understood, and even appreciated, and a caring developed that showed itself in many ways. If things went wrong for someone the concern was genuine and profound. I came to value the guidance I would be given by the crew about their mates. It was never offered in a spirit of gossip but in the hope that every possible avenue of practical support was being explored. There were times, not too often, when the information offered had become distorted somewhere along the line, and when acted upon could result in a minefield of misunderstanding. I once visited a lady who, I was led to believe, was a mother bereaved as a result of an offshore accident. My doorstep conversation revealed her to be unmarried and childless. The dialogue was of such delicacy that the true state of affairs took quite a long time to emerge. I was then directed to a sister (familial communication was not a strong point). This time it turned out she had no offshore connections, but a visit to yet another sister turned out to be where I should have been in the first place.

Offshore, my powers of observation of my fellow men and women improved enormously. The perfunctory glance was simply not good enough and, as ever, it was the eyes that told me so much. They could reflect a welter of emotions, sometimes with a wariness as well as a weariness. Often they seemed far too old for their young faces. The strain was there but so too was the humour – such an effective camouflage. Joking and leg-pulling were features of offshore life but the transformation was absolute if the alarm sounded. Immediately there was a total change and in an instant a very professional crew would be waiting to see what might be required of them. The rubicon had been crossed with the destruction of Piper Alpha and from that time crews accepted that the seemingly impossible was possible – their eyes showed this to be true.

Working offshore is an activity for the younger man. I began to realise I was not so young, as once again I wormed myself into a survival suit for a short inter-platform flight. Wriggling one's arms through the tight wristbands – useless if they were not a snug fit – and getting ones legs into unyielding trousers could be surprisingly tiring, particularly if one went through the performance several times a day. I felt my dotage was confirmed when confronted with the rope to haul myself into the small passenger section of the Bell helicopter. I realised I would have to call it a day when I found myself going through the indignity of hoisting myself on board on my hands and knees. No longer was I young and lithe, swinging, tarzan-like, into a passenger seat! Confirmation that I was really over the hill came with the discovery that on some platforms an unofficial mentor had been appointed who would keep an eye on the old man of the North

Sea and make sure he reached his muster station in a time of emergency. For long enough I had felt that with so many platforms, and different drills to remember, the occasional muddle was excusable, and hopefully, forgivable. But increasingly I found myself less convinced by my own excuses.

And, there were other incidents. I made my mark (an understatement) when clearing up after Christmas Dinner and being so generous with the soap powder in the galley's vast industrial dish washer, that it could not believe its luck and did its best to make the kitchen premises disappear under what looked like a snow drift. Neither was I the flavour of the month when my portable clock, with a distinctive alarm, went off in my baggage and until I could be found the whole platform was put on alert. This was not earth-shaking but in my eyes such incidents were sufficiently significant to make me accept that there is no room for anyone offshore who can become a liability. 'Safety always takes priority', was framed above the desk of one OIM. With great reluctance I realised it was time to make a reasonably dignified exit before I compromised that priority.

The one great plus in my offshore life was that it gave me, at times, opportunities for reflection. It was the breathing space that helped me sort out the difference between busyness and productivity. This was especially valuable when you were involved in an activity having no quantifiable end product. It is possible to become almost paranoid in looking for signs of achievement. Learning to live with the knowledge you had done your best – with little indication of what that 'best' might be – required considerable self-discipline.

There came a time when I had to spend a couple of weeks in hospital. There was an unending procession of visitors to the bedside and flowers arrived in such profusion that the ward began to look alarmingly like a funeral parlour. Happily I was not so ill that I could not value the tangible expressions of goodwill. Thank goodness for flowers and fruit and magazines. They can convey feelings when speeches are not a part of our life style.

At different times my duties had taken me, quite literally, the length and breadth of the United Kingdom, but a major incident in the Far East had resulted in a considerable loss of life on the platform and the production company felt my presence there might be helpful. Time was of the essence and I was rushed to the offshore medical centre for a comprehensive dose of inoculations. It seemed to me that when in doubt the medical officer acted on the principle of 'better safe than sorry'. I am sure I would have had every reason to be grateful to have become a human pin-cushion. However, there was a snag. A few days later, still feeling bruised and battered, I was told that the trip had been cancelled. The good thing was that this non-event set me thinking about prioritising my work.

It can be all too easy to be so accommodating that before long work ceases to have an identifiable pattern with obvious parameters. The danger of spreading yourself too thin becomes apparent. The time had come to identify the main objectives of the Chaplaincy and draw a ring round them.

I suppose in some cases job definition is not difficult because the work is self-defining, but a chaplaincy can be amorphous and the individual has to work out the demarcation lines as he becomes familiar with all that is involved.

My own great problem is a capacity for self-justification which comes into play when I get myself involved in something that falls outside my own specific remit. I was acutely aware that there was no room for a North Sea dilettante – if such an individual can be imagined. The years of involvement had made everything quite clear and the objectives could be prioritised. There was the spiritual leadership and pastoral care of those offshore, together with support and concern for those on the beach who made the decisions, formulated policy and implemented change. Understanding relationships had to be established with UKOOA and the Energy Institute. The former were the paymasters and the latter lent perspective, and were a constant reminder of the global operation in which the UK was a playing partner. The government of the day had to be reminded at all times of the human implications of every pronouncement uttered and decision taken relating to the oil industry. Finally, through talks, broadcasts, articles, seminars and speeches, the priority of people in the whole offshore operation had to be underlined. The industry was dedicated to the extraction of fossil fuels but somehow, somewhere, someone had to point up the people element, without which the dedication would be meaningless, the extraction impossible, and the slick phrases in the glossy annual reports paying tribute to the offshore workforce would be total humbug.

14

Still Developing

LOG EXCERPTS

⌕ *Colin Jones forecast that eventually the North Sea will be full of skilled labour from the Far East. Question: Will industrial mission in the UK eventually have its own ayatollah?*

⌕ *The North Sea is full of experienced men who have discovered death.*

'At least you made some of the Christians come out of the woodwork.'

Crew member at an offshore retirement party

The few years of involvement with the offshore oil industry were the most fulfilling in my years as an ordained minister. Other paths of a varied career have been enjoyable, but it was only towards the end of my working life that a clear pattern and a sense of real purpose emerged. Hindsight can have its merits, not least when a backward glance shows how a clear shape was emerging in the development of both my thinking and doing. My school reports testify to a late developer, and I have had to suffer this condition throughout my life. A quirk of temperament makes me inclined to 'hasten slowly' – at least in projects for which there has been no precedent. It has taken me a long time to weigh up the value and purpose of so many of the 'soft' issues which have occupied my life.

In the exercise of ministry there is rarely a quantifiable end product. I have been fortunate to have been fully involved in establishing an international school, creating a new building, developing an old one and constructing a chapel; they are all the tangible results of some sort of visionary impulse but are hardly likely to satisfy the enquirer who asks, 'What do you do?' There will be those who can accept the value of having someone conveniently on tap for baptisms, weddings and funerals. The more I mull over the whole question of the meaning and purpose of the life I chose, and the more I see those times when the role of chaplain, apparently diminished to that of a functionary on the periphery of society, the more, paradoxically, I see the true significance of chaplaincy beginning to emerge.

In my role I faced the occupational hazard of saying Grace at public dinners. At these functions I would find myself perched on the left or right wing of what could be a very long top table. There was only one person to talk to, if he or she was not talking to the person the other side. Usually there was lots of time to scan the tables of hungry, and thirsty, people and realise that you were a part of the throng and yet somehow apart from it. The apparent pointlessness of just being there, be it at a dinner or offshore, is the paradox of chaplaincy. Being present; doing one's own thing; being wonderfully irrelevant; not being result-driven; liberated from career opportunism; yet at the same time confronting all the values of the world and within it offering a pool of stillness, is surely what chaplaincy is all about?

There were many times when I questioned the value of the Chaplaincy, and I would return to my cabin feeling strangely alone, and wondering if I had got it wrong with my passionate belief that the Incarnation with its reconciliation of man with God really was a truth for everyone to share. Maybe, I would ponder, it was a truth which could only make sense within the parameters of institutional religion, and people had to be brought within the fold before the significance of the extraordinary gesture of God to humankind, his only Son, would be understood. Then something would happen, a chance remark, a thoughtful gesture, and I would find I was turning this negative thinking on its head. I came to see that institutional religion can only be personally meaningful when God's entry into human history is made real in the work situation which fills the major part of everyone's active life and with which we are familiar and where we are most at ease.

From the Wendy House to the Directors' Dinner, the chaplain's responsibilities are not theirs but there is one most precious thing he has in common – a shared humanity in all its bewildering diversity. Just by being there, another dimension is added to the mix. The meaning is found in its existence and not in an end product.

When St John's Chapel (the Oil Chapel) was dedicated in the Kirk of St Nicholas in Aberdeen on 25 June 1990, I was privileged to give the address and tried to express some thoughts on the use to which it could, and should, be put:

> Is not this what life is all about? Laughter and tears, confidence and fear, companionship and emptiness? There is not one of us who escapes the harrowing, or does not enjoy the exhilarating. And the mark of our humanity is how our spirits can endure, survive and soar again. It is in the indwelling of the Spirit of Christ who defeated the last enemy – Death – who knows so well the road we travel, that we shall be healed. That is why this Chapel, this holy place, will be a constant source of calm and resource for peace.

The personal pilgrimage – for that is what my chaplaincy had become – was over. I, personally, was more at ease with my faith.

15

An Epilogue

LOG EXCERPT

Touched down on my old friend the flotel, the Safe Gothia. As usual very crowded, and as usual there was a delay. The Admin Clerk asked who I worked for; I pompously replied, 'God', to which he responded, 'I don't believe in that bastard.' I had unChristian thoughts about re-arranging his features, but wisdom prevailed.

> **'There have been so many friendships formed – adversity is a great welder; time alone will show if some suggestionof Christian infection has begun.'**
>
> Final entry in my log

The chopper rose vertically, performed its customary curtsey, levelled out and headed for Aberdeen. At least that was the announced route: instead it circled and headed back to fly over the Fulmar Platform and the nearby FSU. The crews filled the helidecks waving 'Goodbye!' My eyes were blurred and there was a large lump in my throat. The years offshore had been rich in experience and I would not be returning but those I left behind would always be in my thoughts and prayers. A bumpy spiritual journey had come to a happy conclusion in the middle of the North Sea.

Glossary

Abseiler A job for those with a head for heights. Exposed areas of platform structures can be reached for maintenance and repair using climbing techniques instead of the more usual scaffolding.

Back to back A member of a crew who shares a particular responsibility with a colleague presently on the beach to whom he will hand over at the next crew change.

Beach The mainland from which everyone departs and to which everyone returns.

Bear A member of crew. Often someone on contract. They could be notable for their physical bulk, the grubbiness of their overalls and their tattoos. There could be an ear-ring or two.

Bear Pit Where the Bears play in the tea breaks. Freshly baked cakes and scones are delivered from the galley. Here cigarettes can be lit, the world put to rights, and Chelsea and Manchester United, Rangers and Celtic *et al.* can be minutely dissected.

Bottle The capacity to see things through and overcome natural fears. E.g. 'He showed lots of bottle.'

Bottom line The end product of a particular project that is tangible and can be assessed. In their own work chaplains can have a problem with this one.

Bumped Turning up for a flight only to discover one has to give way; usually for an urgently required piece of machinery – good for humility!

Camp Boss The manager responsible for the smooth running, and hygiene, of the galley and every aspect of the catering. Also for the PLQ and the shop.

Cans The headphones worn by all passengers during a flight so that the pilot has instant communication with all on board.

Chopper A helicopter.

Crane operator The artiste on a platform whose skill and judgement makes possible the lifting and lowering of loads to and from a supply vessel often in foul weather when judgement is crucial.

Crew change Usually men and women work offshore for 14-day periods. The reliefs for different areas of work are staggered with the back to backs flying in on different days each week.

Contractor The company that provides the required skills for the operator. Since offshore operations first began, the balance has swung from almost 100 per cent company involvement to far greater autonomy for the contractors working in long-term relationships.

Commuting The travel pattern of those going to and from their work by road, rail and air uniting for the one part of the journey by helicopter. Workers may sleep on one platform and travel daily to another nearby in a smaller shuttle chopper.

Deck foreman Deck space on a platform is in short supply. Everything has to be hoisted by crane from supply boats and stacked with precision. There is great teamwork between the crane operator and the deck foreman who is a maestro at hand signals.

Deck crew Usually double up as helideck crew. Highly trained. Responsible for the control and safety of all on the helideck. They organise baggage and stow it in the hold. Also responsible for organising the nourishment for chopper crews. The deck crew probably breath more fresh air than the rest of the platform put together.

Derrick man High above the drilling floor, secured by a safety belt, he stacks and releases the lengths of pipe. Within the drilling hierarchy the derrick man has already been a roughneck and a roustabout.

Down stream Oil industry activities associated with forecourt pump sales and the supply of commercial fuel.

Ditched Someone who finds himself or herself, through accident, in the sea in what will inevitably be a life- threatening situation.

Driller A man of vast drilling floor experience who controls the direction and speed of the drill, and who, by touch, can tell what is going on thousands of feet below sea level.

Drilling floor No place for the visitor. The greasy wooden slabs increase the hazard of the heavy and dangerous work of inserting, or extracting, the pipe. Even more remarkable are the exhibitions of balance and teamwork.

Dynamic positioning Computer-controlled jets of water ensure that the supply vessel remains alongside the platform no matter the sea state.

Floater A production unit converted from a rig. The floating platform is secured in position by massive chains and anchors.

FSU Fuel Supply Unit.

Flotel A specially constructed structure that can accommodate up to a thousand workers. A flotel is usually moored alongside the platform that is being renovated. Workers board the platform by a bridge that is automatically raised, with much flashing of lights and ringing of alarms, if the wind and sea reach a certain level.

Golden handcuffs The situation can arise when the offshore worker has had enough and wants to return to the beach for work but finds his financial commitments do not make this possible.

HLO Helideck Landing Officer responsible for all pre- and post-flight activity.

HSE The Health and Safety Executive. The government body from whose Olympian heights all safety directives are issued. It is also home to the offshore inspectors.

Medic A man or woman with a state registered nursing qualification; not infrequently with industrial nursing experience. Many medics have served in the armed forces.

Muster station Everyone on a platform has an assembly point if an alarm sounds.

North Sea Central The location of offshore installations some one hundred miles east of Dundee.

North Sea North The location of offshore installations in the Shetland Basin sixty miles north east of the islands. There are platforms to the North and South of the basin.

North Sea South The smaller gas platforms some twenty miles off the East Anglian Coast.

No show When a member of a relieving crew fails to report at check-in for his next tour offshore.

NRB Not required back. The damning letters that can appear on a employee's record and can make it difficult to obtain further employment offshore. No reason has to be given when an employee's services are no longer required.

OIM The Offshore Installation Manager. The key individual on a platform. He is vested with the same powers as a ship's captain. He is responsible for the safe production of oil and gas, the condition of the platform and the wellbeing of all on board.

Operator Most platforms are funded by syndicates of the major oil producers, one of which will be designated the operator with responsibility for the successful performance of the installation.

Rig The layperson's description of all offshore installations. However, rigs are mobile and designed to explore the oceans in search of commercially viable quantities of oil and gas. It is their work that determines the location of the more permanent platforms.

Platform A permanently sited structure. At first these structures rested on the seabed, but with exploration in deeper waters they have become 'floaters' dynamically positioned. They extract oil and gas in commercially viable quantities.

PLQ Personal Living Quarters. The steel cubes can be up to eight storeys high (there are lifts). The structure will include cabins, bathrooms, laundries, mess rooms, recreational facilities and most probably the radio room, the main offices and the sick bay.

Roustabout and Roughneck Workers on the drilling floor. The twelve-hour shifts involve the continuous movement of heavy material, which is both physically exhausting and requires constant alertness.

Safety officer Responsible for the observance of safety regulations, the maintenance of all safety equipment and the training of its operators. The safety officer also briefs all new arrivals to a platform.

Scaffy Builds, creates is maybe a more accurate word, the structures which enable repairs and maintenance to be safely carried out. The working areas created can be in very exposed locations.

Semi-sub A structure whose hull is on the surface when under way either by tow or its own power. Once on site the hulls are flooded and sink below sea level to generate greater stability.

Sharp-end Where it all happens. The men and women who are directly responsible for the production of oil and gas.

Shuttle The small helicopters that fly between platforms carrying passengers and freight.

Soft side The long established description for all activities that are to do with the human side of industry, that is, human resources, welfare, health and chaplaincy.

Split shift An arrangement between a man or woman and their 'back to back' over the organisation of their work, e.g. who will be offshore at Christmas or New Year. It is not unusual for one person to spend three successive weeks offshore so that their back to back may have an extended vacation. Note: the gesture will be reciprocated.

Stand by vessel The small vessels that lie off all installations at all times, in all weathers. They draw closer to the platforms when helicopters are landing or taking off, or when work is being carried out in exposed situations. They take great pride in their speed of reaction should anyone have the misfortune to be ditched.

Supply vessel Specially designed, high powered vessels with a large unobstructed deck area for the transport of supplies from the beach to the platforms. The supplies travel in containers.

Sumburgh Shetland's main airport with an over-large terminal built somewhat optimistically to accommodate projected air traffic that never materialised. At one time many workers going to the Shetland Basin flew here fixed- wing from Aberdeen. The heliport is the main centre of activity.

Scatsta A more primitive landing strip serving the same purpose as Sumburgh.

Tiger See 'Bear' – same animal, different name. Also one of the North Sea workhorses – a variant of the Super Puma helicopter.

Trip The designated time for a working period offshore – normally fourteen days.

Upstream All aspects of the oil industry associated with exploration, assessment and recovery.

Weather window At certain times of the year it can be reasonably assumed that the sea state will be sufficiently calm to allow the towing of the modules that together make up a platform, and to permit their very precise installation. Presumably the reverse procedure will take place when decommissioning takes place.

Wendy House A (slightly) refined version of the Bear Pit.

North Sea Oil Fields

Unst

Sulum Voe & Scatsta

Shetlands

Sumburgh Head

Orkney

Kirkwall

Flotta

Peterhead

Aberdeen

Scotland

Dundee

1
3
2
5
4
8
9
12
13
6
7
10
11
14
15
16

17

19
18
20

21

22

23

24

25

26

27